Lecture Notes in Mathematics 1564

Editors:
A. Dold, Heidelberg
B. Eckmann, Zürich
F. Takens, Groningen

Subseries:
Mathematisches Institut der Universität
und Max-Planck-Institut für Mathematik,
Bonn - vol. 18

Advisor:
F. Hirzebruch

Jay Jorgenson   Serge Lang

Basic Analysis of Regularized Series and Products

Springer-Verlag
Berlin Heidelberg New York
London Paris Tokyo
Hong Kong Barcelona
Budapest

Authors

Jay A. Jorgenson
Serge Lang
Department of Mathematics
Yale University
Box 2155 Yale Station
New Haven, CT 06520, USA

Mathematics Subject Classification (1991): 11M35, 11M41, 11M99, 30B50, 30D15, 35P99, 35S99, 39B99, 42A99

(Authors' Note: there is no MSC number for regularized products, but there should be.)

ISBN 3-540-57488-3 Springer-Verlag Berlin Heidelberg New York
ISBN 0-387-57488-3 Springer-Verlag New York Berlin Heidelberg

2146/3140-543210 - Printed on acid-free paper

Foreword

The two papers contained in this volume provide results on which a series of subsequent papers will be based, starting with [JoL 92b], [JoL 92d] and [JoL 93]. Each of the two papers contains an introduction dealing at greater length with the mathematics involved.

The two papers were first submitted in 1992 for publication in *J. reine angew. Math.* A referee emitted the opinion: "While such generalized products are of interest, they are not of such central interest as to justify a series of long papers in expensive journals." The referee was cautious, stating that this "view is subjective", and adding that he "will leave to the judgement of the editors whether to pass on this recommendation to the authors". The recommendation, in addition not to publish "in expensive journals", urged us to publish a monograph instead. In any case, the editors took full responsibility for the opinion about the publication of our series "in expensive journals". We disagree very strongly with this opinion. In fact, one of the applications of the complex analytic properties of regularized products contained in our first paper is to a generalization of Cramér's theorem, which we prove in great generality, and which appears in *Math. Annalen* [JoL 92b]. The referee for *Math. Ann.* characterized this result as "important and basic in the field".

Our papers were written in a self-contained way, to provide a suitable background for an open-ended series. Thus we always considered the possible alternative to put them in a Springer Lecture Note, and we are very grateful to the SLN editors and Springer for publishing them.

Acknowledgement: During the preparation of these papers, the first author received support from the NSF Postdoctoral Fellowship DMS-89-05661 and from NSF grant DMS-93-07023. The second author benefited from his visits at the Max Planck Institut in Bonn.

Part I

Some Complex Analytic Properties of Regularized Products and Series

Content

Part II

A Parseval Formula for Functions with a Singular Asymptotic Expansion at the Origin

Content

Part I

Some Complex Analytic Properties of Regularized Products and Series

Introduction

We shall describe how parts of analytic number theory and parts of the spectral theory of certain operators (differential, pseudo-differential, elliptic, etc.) are being merged under a more general analytic theory of regularized products of certain sequences satisfying a few basic axioms. The most basic examples consist of the sequence of natural numbers, the sequence of zeros with positive imaginary part of the Riemann zeta function, and the sequence of eigenvalues, say of a positive Laplacian on a compact manifold. The resulting theory applies to the zeta and L-functions of number theory, or representation theory and modular forms, to Selberg-like zeta functions in spectral theory, and to the theory of regularized determinants familiar in physics and other parts of mathematics.

Let $\{\lambda_k\}$ be a sequence of distinct complex numbers, tending to infinity in a sector contained in the right half plane. We always put $\lambda_0 = 0$ and $\lambda_k \neq 0$ for $k \geq 1$. We are also given a sequence $\{a_k\}$ of complex numbers. We assume the routine conditions that the Dirichlet series

$$\sum_{k=1}^{\infty} \frac{a_k}{\lambda_k^{\sigma}} \quad \text{and} \quad \sum_{k=1}^{\infty} \frac{1}{\lambda_k^{\sigma}}$$

converge absolutely for some $\sigma > 0$. If $a_k \in \mathbf{Z}_{\geq 0}$ for all k, we view a_k as a multiplicity of λ_k, and we call this **the spectral case**. We may form other functions, namely:

The **theta series** $\theta(t) = \sum_{k=0}^{\infty} a_k e^{-\lambda_k t}$ for $t > 0$;

The **zeta function** $\zeta(s) = \sum_{k=1}^{\infty} a_k \lambda_k^{-s}$;

The **Hurwitz zeta function** $\zeta(s, z) = \sum_{k=0}^{\infty} a_k (z + \lambda_k)^{-s}$;

The **xi function** $\xi(s, z) = \Gamma(s)\zeta(s, z)$, which can be written as the **Laplace-Mellin transform** of the theta function, that is

$$\xi(s,z) = \int_0^\infty \theta(t) e^{-zt} t^s \frac{dt}{t} = \mathbf{LM}\theta(s,z).$$

For the sequence $\{\lambda_k\}$ with $a_k = 1$ for all k, consider the derivative

$$\zeta'(s) = \sum_{k=1}^{\infty} \frac{-\log \lambda_k}{\lambda_k^s}.$$

Putting $s = 0$ formally, as Euler would do (cf. [Ha 49]), we find

$$\zeta'(0) = \sum_{k=1}^{\infty} -\log \lambda_k.$$

Therefore if the zeta function has an analytic continuation at $s = 0$, then

$$\exp(-\zeta'(0)) = \prod_{k=1}^{\infty} \lambda_k$$

may be viewed as giving a value for the meaningless infinite product on the right. Similarly, using the sequence $\{\lambda_k + z\}$ instead of $\{\lambda_k\}$, we would obtain a value for the meaningless infinite product

$$\mathbf{D}(z) = \exp(-\zeta'(0,z)) = \prod_{k=1}^{\infty} (\lambda_k + z),$$

where the derivative here is the partial derivative with respect to the variable s. To make sense of this procedure in the spectral case, under certain conditions one shows that the sequence $\{\lambda_k\}$ also determines:

The **regularized product**

$$D(z) = e^{P(z)} E(z),$$

where P is a normalizing polynomial, and $E(z)$ is a standard Weierstrass product having zeros at the numbers $-\lambda_k$ with multiplicity a_k. The degree of P and the order of the Weierstrass product will be characterized explicitly below in terms of the sequence $\{\lambda_k\}$ and appropriate conditions.

We keep in mind the following four basic examples of the spectral case.

Example 1. The gamma function. Let $\lambda_k = k$ range over the natural numbers. Then the zeta function is the Riemann zeta function $\zeta_{\mathbf{Q}}$, and the Hurwitz zeta function is the classical one (whence the name we have given in the general case). The theta function is simply

$$\theta(t) = \sum_{k=0}^{\infty} e^{-kt} = \frac{1}{1-e^{-t}}.$$

The corresponding Weierstrass product is that of the gamma function.

Example 2. The Dedekind zeta function. Let F be an algebraic number field. The **Dedekind zeta function** is defined for $\mathrm{Re}(s) > 1$ by the series

$$\zeta_F(s) = \sum \mathbf{N}\mathfrak{a}^{-s}$$

where $\mathfrak{a}$ ranges over the (non-zero) ideals of the ring of algebraic integers of F, and $\mathbf{N}\mathfrak{a}$ is the absolute norm, in other words, the index $\mathbf{N}\mathfrak{a} = (\mathfrak{o} : \mathfrak{a})$. Then

$$\theta(t) = \sum_{k=0}^{\infty} a_k e^{-kt}$$

where a_k is the number of ideals $\mathfrak{a}$ such that $\mathbf{N}\mathfrak{a} = k$. This theta function is different from the one which occurs in Hecke's classical proof of the functional equation of the Dedekind zeta functions (cf. [La 70], Chapter XIII). Of course, this example extends in a natural way to Dirichlet, Hecke, Artin, and other L-series classically associated to number fields. In these extensions, a_k is usually not an integer.

Zeta functions arising from representation theory and the theory of automorphic functions constitute an extension of the present example, but we omit here further mention of them to avoid having to elaborate on their more complicated definitions.

Example 3. Regularized determinant of an operator. In this case, we let $\{\lambda_k\}$ be the sequence of eigenvalues of an operator. In the most classical case, the operator is the positive Laplacian on a compact Riemannian manifold, but other much more complicated examples also arise naturally, involving possibly non-compact manifolds or pseudo differential operators. Suitably normalized, the function $D(z)$ is viewed as a regularized determinant (generalizing the characteristic polynomial in finite dimensions).

Example 4. Zeros of the zeta function. Let $\{\rho_k\}$ range over the zeros of the Riemann zeta function with positive imaginary part. Let a_k be the multiplicity of ρ_k, conjecturally equal to 1. Put $\lambda'_k = \rho_k/i$. The sequence $\{\lambda'_k\}$ is thus obtained by rotating the vertical strip to the right, so that it becomes a horizontal strip. A theorem of Cramér [Cr 19] gives a meromorphic continuation (with a logarithmic singularity at the origin) for the function

$$2\pi i V(z) = \sum a_k e^{\rho_k z} = \sum a_k e^{-\lambda'_k t}$$

which amounts to a theta function in this case (after the change of variables $z = it$).

This fourth example generalizes as follows. Suppose given a sequence $\{\lambda_k\}$ such that the corresponding zeta function has an Euler product and functional equation whose fudge factors are of regularized product type. (These notions will be defined quite generally in [JoL 92b].) We are then led to consider the sequence $\{\lambda'_k\}$ defined as above. From §7, one sees that a regularized product exists for the sequence $\{\lambda_k\}$. We will show in [JoL 92b] that a regularized product also exists for the sequence $\{\lambda'_k\}$. Passing from $\{\lambda_k\}$ to $\{\lambda'_k\}$ will be called **climbing the ladder** in the hierarchy of regularized products.

Basic Formulas. The example of the gamma function provides a basic table of properties which can be formulated and proved under some additional basic conditions which we shall list in a moment. The table includes:

The multiplication formula
The Lerch formula
The (other) Gauss formula
The Stirling formula
The Hankel formula
The Mellin inversion formula
The Parseval formula.

The multiplication formula may be viewed as a special case of the Artin formalism treated in [JoL 92d]. The Parseval formula, which determines the Fourier transform of Γ'/Γ as a distribution on a vertical line will be addressed in the context of a general result in Fourier analysis in [JoL 92c]. Here we show that the other formulas can be expressed and proved in a general context, under certain axioms (covering all four examples and many more complicated analogues). We shall find systematically how the simple expression $1/(1-e^{-t})$ is replaced by theta functions throughout the formalism developed in this part. More generally, whenever the above expression occurs in mathematics, one should be on the lookout for a similar more general structure involving a theta function associated to a sequence having a regularized product.

The Asymptotic Expansion Axiom. The main axiom is a certain asymptotic expansion of the theta function at the origin, given as **AS 2** in §1; namely, we assume that there exists a sequence of complex numbers $\{p\}$ whose real parts tend to infinity, and polynomials B_p such that

$$\theta(t) \sim \sum_p B_p(\log t)t^p.$$

The presence of log terms is essential for some applications.

In Example 1, the asymptotic expansion of the theta function is immediate, since $\theta(t) = 1/(1-e^{-t})$. In Example 2, this expansion follows from the consideration of §7 of the present part. In both Examples 1 and 2, B_p is constant for all p, so we say that there are no log terms.

In the spectral theory of Example 3, Minakshisundaram-Pleijel [MP 49] introduced the zeta function formed with the sequence of eigenvalues λ_k, and Ray-Singer introduced the so-called analytic torsion [RS 73], namely $\zeta'(0)$. Voros [Vo 87] and Cartier-Voros [CaV 90] gave further examples and results, dealing with a sequence of numbers whose real part tends to infinity. We have found their axiomatization concerning the corresponding theta function useful. However, both articles [Vo 87] and [CaV 90] leave some basic questions open in laying down the foundations of regularized products. Voros himself states: "In the present work, we shall not be concerned with rigorous proofs, which certainly imply additional regularity properties for the sequence $\{\lambda_k\}$." Furthermore, Cartier-Voros have only certain specific and special applications in mind (the Poisson summation formula and the Selberg trace formula in the case of compact Riemann surfaces). Because of our more general asymptotic expansion for the theta function, the theory becomes applicable to arbitrary compact manifolds with arbitrary Riemannian metrics and elliptic pseudo-differential operators where the log terms appear starting with [DuG 75], and continuing with [BrS 85], [Gr 86] and [Ku 88] for the spectral theory. In this case, the theta function is the trace of the heat kernel, and its asymptotic expansion is proved as a consequence of an asymptotic expansion for the heat kernel itself.

In Example 4 for the Riemann zeta function, the asymptotic expansion follows as a corollary of Cramér's theorem. A log term appears in the expansion. The generalization in [JoL 92b] involves some extra work. Some of the arguments used to prove the asymptotic expansion **AS 2** are given in §5 of the present part (especially Theorem 5.11), because they are directly related to those used to prove Stirling's formula. Indeed, the Stirling formula gives an asymptotic expansion for the log of a regularized product at infinity; in [JoL 92b] we require in addition an asymptotic expansion for the Laplace transform of the log of the regularized product in a neighborhood of zero.

Normalization of the Weierstrass Product by the Lerch Formula. We may now return to describe more accurately our normalization of the Weierstrass product. When the Hurwitz zeta function $\zeta(s, z)$ is holomorphic at $s = 0$, and all numbers a_k are positive integers, there is a unique entire functions $\mathbf{D}(z)$ whose zeros are the numbers $-\lambda_k$ with multiplicities a_k, with a normalized Weierstrass product such that the **Lerch formula** is valid, namely

$$\log \mathbf{D}(z) = -\zeta'(0, z),$$

where the derivative on the right is with respect to the variable s.

As to the Weierstrass order of $\mathbf{D}$, let p_0 be a leading exponent in the asymptotic expansion for $\theta(t)$, i.e. $\operatorname{Re}(p_0) \leq \operatorname{Re}(p)$ for all p such that $B_p \neq 0$. Let M be the largest integer $< -\operatorname{Re}(p_0)$. Then $M + 1$ is the order of $\mathbf{D}$.

In the general case with the log terms present in the asymptotic expansion, the Hurwitz zeta function $\zeta(s, z)$ may be meromorphic at $s = 0$ instead of being holomorphic, and we show in §3 how to make the appropriate definitions so that a similar formula is valid.

In Example 1, this formula is the classical Lerch formula. In Example 2, and various generalizations to L-functions of various types, the formula is new as far as we know. In Example 3, the formula occurs in many special cases of the theory of analytic torsion of Ray-Singer and in Voros [Vo 87], formula (4.1). In Example 4, the formula specializes to a formula discovered by Deninger for the Riemann zeta function $\zeta_{\mathbf{Q}}$ (see [De 92], Theorem 3.3).

Applications. Aside from developing a formalism which we find interesting for its own sake, we also give systematically fundamental analytic results which are used in the subsequent series of parts, including not only the generalization of Cramér's theorem mentioned above, but for instance our formulation of general explicit formulas analogous to those of analytic number theory (see [JoL 93]). These particular applications deal with cases when the zeta function has an Euler product and functional equation. Such cases may arise from a regularized product by a change of variables $z = s(s-1)$. The Selberg zeta function itself falls in this category. However, so far the Euler product does not play a role. A premature change of variables $z = s(s-1)$ obscures the basic properties of the regularized product and Dirichlet series which do not depend on the Euler product.

Although, as we have pointed out, some special cases of our formulas are known, many others are new. Our results and formulas concerning regularized products are proved in sufficient generality to apply in several areas of mathematics where zeta functions occur, e.g. analytic number theory, representation theory, spectral theory, ergodic theory and dynamical systems, etc. For example, our Lerch formula is seen to apply to Selberg type zeta functions not only for Riemann surfaces but for certain higher dimensional manifolds as well.

Furthermore, our general principle of climbing the ladder of regularized products applies to the scattering determinant associated to a non-compact hyperbolic Riemann surface of finite volume. In [JoL 92b] we shall use results of Selberg to show that the scattering determinant satisfies our axioms, and hence is of regularized product type. As a second application of our Cramér's theorem, we then conclude that the Selberg zeta function in the non-compact case is also of regularized product type. These facts were not known previously.

Our theory also applies to Ruelle type zeta functions arising in ergodic theory and dynamical systems (for example, see [Fr 86] and references given in that part).

Therefore, we feel that it is timely to deal systematically with the theory of regularized products, which we find central in mathematics.

For the convenience of the reader, a table of notation is included at the end of this part.

§1. Laplace-Mellin Transforms

We first recall some standard results concerning Laplace-Mellin transforms. The **Mellin transform** of a measurable function f on $(0, \infty)$ is defined by

$$\mathbf{M}f(s) = \int_0^\infty f(t)t^s \frac{dt}{t}.$$

The **Laplace transform** is defined to be

$$\mathbf{L}f(z) = \int_0^\infty f(t)e^{-zt} \frac{dt}{t}.$$

The **Laplace-Mellin transform** combines both, with the definition

$$\mathbf{LM}f(s, z) = \int_0^\infty f(t)e^{-zt}t^s \frac{dt}{t}.$$

We now worry about the convergence conditions. The next lemma is standard and elementary.

Lemma 1.1. *Let I be an interval of real numbers, possibly infinite. Let U be an open set of complex numbers, and let $f = f(t, z)$ be a continuous function on $I \times U$. Assume:*

(a) *For each compact subset K of U the integral*

$$\int_I f(t, z)\, dt$$

is uniformly convergent for $z \in K$.

(b) *For each t the function $z \mapsto f(t, z)$ is holomorphic.*

Let

$$F(z) = \int_I f(t, z)\, dt.$$

Then the second partial $\partial_2 f$ satisfies the same two conditions as f, the function F is holomorphic on U, and

$$F'(z) = \int_I \partial_2 f(t, z)\, dt.$$

For a proof, see [La 85], Chapter XII, §1. We then have immediately:

Lemma 1.2. Special Case. *Let z be such that* $\mathrm{Re}(z) > 0$ *and let* $b_p \in \mathbf{C}$*. Then for* $\mathrm{Re}(s) > 0$ *we have*

$$\int_0^\infty b_p e^{-zt} t^s \frac{dt}{t} = b_p \frac{\Gamma(s)}{z^s},$$

the integral being absolutely convergent, uniformly for

$$\mathrm{Re}(z) \geq \delta_1 > 0 \quad \text{and} \quad \mathrm{Re}(s) \geq \delta_2 > 0.$$

General Case. *For any polynomial B, let $B(\partial_s)$ be the associated constant coefficient differential operator. For* $\mathrm{Re}(z) > 0$, $p \in \mathbf{C}$ *and* $\mathrm{Re}(s+p) > 0$ *we have*

$$\int_0^\infty e^{-zt} B(\log t) t^{s+p} \frac{dt}{t} = B(\partial_s) \left[\frac{\Gamma(s+p)}{z^{s+p}} \right].$$

The proof of Lemma 1.2 follows directly from an interchange of differentiation and integration, which is valid for z and s in the above stated region.

At this point let us record several very useful formulas. For any z with $\mathrm{Re}(z) > 0$ and any s with $\mathrm{Re}(s) > 0$,

$$\int_1^\infty e^{-zt} t^s \frac{dt}{t} = \frac{1}{z^s} \int_{1/z}^\infty e^{-u} u^s \frac{du}{u} = \frac{\Gamma(s)}{z^s} - \int_0^1 e^{-zt} t^s \frac{dt}{t}$$

where the path of integration in the second integral is such that u/z is real. By expanding e^{-zt} in a power series about the origin, we have

$$\int_0^1 e^{-zt} t^s \frac{dt}{t} = \sum_{k=0}^\infty \frac{(-z)^k}{k!} \frac{1}{s+k},$$

which shows that the given integral can be meromorphically continued to all $s \in \mathbf{C}$ and all $z \in \mathbf{C}$. The integration by parts formula

$$s \int_0^1 e^{-zt} t^s \frac{dt}{t} = e^{-z} + z \int_0^1 e^{-zt} t^{s+1} \frac{dt}{t},$$

also provides a meromorphic continuation of the given integral.

The next lemma shows how an asymptotic expression for a function $f(t)$ near $t = 0$ gives a meromorphic continuation and an asymptotic expansion at infinity for the Laplace-Mellin transform of $f(t)$. We first consider a special case.

Lemma 1.3. Special Case. *Let f be piecewise continuous on $(0, \infty)$. Assume:*

(a) *$f(t)$ is bounded for $t \to \infty$.*

(b) *$f(t) = b_p t^p + O(t^{\mathrm{Re}(q)})$ for some $b_p \in \mathbf{C}$, $p, q \in \mathbf{C}$ such that $\mathrm{Re}(p) < \mathrm{Re}(q)$, and $t \to 0$.*

Then for $\mathrm{Re}(s) = \sigma > -\mathrm{Re}(p)$ and $\mathrm{Re}(z) > 0$ the Laplace-Mellin integral

$$\mathbf{LM}f(s,z) = \int_0^\infty f(t) e^{-zt}\, t^s\, \frac{dt}{t}$$

converges absolutely, and for $\mathrm{Re}(s) > -\mathrm{Re}(q)$ the function $\mathbf{LM}f$ has a meromorphic continuation such that

$$\mathbf{LM}f(s,z) = b_p \frac{\Gamma(s+p)}{z^{s+p}} + g(s,z)$$

where for fixed z, $s \mapsto g(s,z)$ is holomorphic for $\mathrm{Re}(s) > -\mathrm{Re}(q)$. The only possible poles in s of $\mathbf{LM}f$ when $\mathrm{Re}(z) > 0$ and when $\mathrm{Re}(s+q) > 0$ are at $s = -p - n$ with $n \in \mathbf{Z}_{\geq 0}$. All poles are simple, and the residue at $s = -p$ is b_p.

Proof. We decompose the integral into a sum:

$$\mathbf{LM}f(s,z) = \int_0^\infty (f(t) - b_p t^p) e^{-zt}\, t^s\, \frac{dt}{t} + \int_0^\infty e^{-zt} b_p t^{s+p}\, \frac{dt}{t}$$

$$= \int_0^1 (f(t) - b_p t^p) e^{-zt}\, t^s\, \frac{dt}{t}$$

$$+ \int_1^\infty (f(t) - b_p t^p) e^{-zt}\, t^s\, \frac{dt}{t} + b_p \frac{\Gamma(s+p)}{z^{s+p}}.$$

The second integral, from 1 to ∞, is entire in s, and converges uniformly for all z such that $\mathrm{Re}(z) \geq \delta > 0$, and for all s such that $\mathrm{Re}(s)$ is in a finite interval of $\mathbf{R}$. Also, we have

$$b_p \frac{\Gamma(s+p)}{z^{s+p}} = \frac{b_p}{s+p} - b_p(\gamma + \log z) + O(s+p),$$

since

$$\Gamma(s) = \frac{1}{s} - \gamma + O(s).$$

So there remains to analyze the first integral, from 0 to 1. More generally, let $q \in \mathbf{C}$, and let g be piecewise continuous on $(0, 1]$ satisfying

$$g(t) = O(t^{\mathrm{Re}(q)}) \quad \text{for } t \to 0.$$

Then the integral

$$I_1(s, z) = \int_0^1 g(t) e^{-zt} t^s \frac{dt}{t}.$$

is obviously holomorphic in $z \in \mathbf{C}$ and $\mathrm{Re}(s) > -\mathrm{Re}(q)$, by Lemma 1.1. This concludes the proof of Lemma 1.3. $\square$

We are now going to show the effect of introducing logarithmic terms. For $p \in \mathbf{C}$, we let B_p denote a polynomial with complex coefficients and we put

$$b_p(t) = B_p(\log t).$$

We then let $B(\partial_s)$ be the corresponding constant coefficient partial differential operator.

Lemma 1.3. General Case. *Let f be piecewise continuous on $(0, \infty)$. Assume:*

(a) *$f(t)$ is bounded for $t \to \infty$.*

(b) *$f(t) = b_p(t)t^p + O(t^{\mathrm{Re}(q)}|\log t|^m)$ for some function*

$$b_p(t) = B_p(\log t) \in \mathbf{C}[\log t],$$

such that $\mathrm{Re}(p) < \mathrm{Re}(q)$, $m \in \mathbf{Z}_{\geq 0}$, and $t \to 0$.

Then for $\mathrm{Re}(s) = \sigma > -\mathrm{Re}(p)$ *and* $\mathrm{Re}(z) > 0$ *the Laplace-Mellin integral*

$$\mathbf{LM}f(s,z) = \int_0^\infty f(t)e^{-zt}\, t^s \,\frac{dt}{t}$$

converges absolutely, and for $\mathrm{Re}(s) > -\mathrm{Re}(q)$ *the function* $\mathbf{LM}f$ *has a meromorphic continuation such that*

$$\mathbf{LM}f(s,z) = B_p(\partial_s)\left[\frac{\Gamma(s+p)}{z^{s+p}}\right] + g(s,z)$$

where for fixed z, $s \mapsto g(s,z)$ *is holomorphic for* $\mathrm{Re}(s) > -\mathrm{Re}(q)$. *The only possible singularity of* $\mathbf{LM}f$ *when* $\mathrm{Re}(z) > 0$ *and* $\mathrm{Re}(s+q) > 0$ *are poles of order at most* $\deg B_p + 1$ *at* $s \doteq -p-n$ *with* $n \in \mathbf{Z}_{\geq 0}$.

The proof is the same as in the special case invoking Lemma 1.1.

In brief, the presence of the logarithmic term $b_p(t)$ introduces a pole at $s = -p$ of order $\deg B_p + 1$, and Lemma 1.3 explicitly describes the polar part of the expansion of the Laplace-Mellin integral near $s = -p$.

The preceding lemmas give us information for

$$g(t) = O(t^{\mathrm{Re}(q)}|\log t|^m).$$

We end our sequence of lemmas by describing more precisely the behavior due to the term $b_p(t)t^p$.

Lemma 1.4. *Let* f *be piecewise continuous on* $(0, \infty)$. *Assume:*

(a) *There is some* $c \in \mathbf{R}$ *such that* $f(t) = O(e^{ct})$ *for* $t \to \infty$.

(b) $f(t) = b_p(t)t^p + O(t^{\mathrm{Re}(q)}|\log t|^m)$ *with* $b_p(t) = B_p(\log t)$, $\mathrm{Re}(p) < \mathrm{Re}(q)$, $m \in \mathbf{Z}_{\geq 0}$ *and* $t \to 0$.

Then $\mathbf{LM}f(s,z)$ *is meromorphic in each variable for*

$$\mathrm{Re}(z) > c \quad \text{and} \quad \mathrm{Re}(s) > -\mathrm{Re}(q)$$

except possibly for poles at $s = -(p+n)$ *with* $n \in \mathbf{Z}_{\geq 0}$ *of order at most* $\deg B_p + 1$.

Proof. We split the integral:

$$\int_0^\infty f(t)e^{-zt}t^s \frac{dt}{t} = I_1(s,p,z) + I_2(s,p,z) + I_3(s,z)$$

where

$$I_1(s,p,z) = \int_0^1 (f(t) - b_p(t)t^p)e^{-zt}t^s \frac{dt}{t}, \tag{1}$$

$$I_2(s,p,z) = \int_0^1 B_p(\log t)e^{-zt}t^{s+p} \frac{dt}{t}, \tag{2}$$

$$I_3(s,z) = \int_1^\infty f(t)e^{-zt}t^s \frac{dt}{t}. \tag{3}$$

The specified poles in s are going to come only from I_2. That is:

$I_1(s,p,z)$ is holomorphic for $z \in \mathbf{C}$ and $\mathrm{Re}(s) > -\mathrm{Re}(q)$.

$I_2(s,p,z)$ has a meromorphic continuation given by expanding e^{-zt} in its Taylor series and integrating term by term to get

$$I_2(s,p,z) = \sum_{n=0}^{\infty} (-1)^n \frac{z^n}{n!} B_p(\partial_s)\left[\frac{1}{n+p+s}\right].$$

$I_3(s,z)$ is holomorphic for $\mathrm{Re}(z) > c$ and all $s \in \mathbf{C}$. The theorem follows. □

We now consider an infinite sequence $L = \{\lambda_k\}$ of distinct complex numbers satisfying:

DIR 1. For every positive real number c, there is only a finite number of k such that $\mathrm{Re}(\lambda_k) \leq c$.

We use the convention that $\lambda_0 = 0$ and $\lambda_k \neq 0$ for $k \geq 1$. Under condition **DIR 1** we delete from the complex plane $\mathbf{C}$ the horizontal half lines going from $-\infty$ to $-\lambda_k$ for each k, together, when necessary, the horizontal half line going from $-\infty$ to 0. We define the open set:

$\mathbf{U}_L$ = the complement of the above half lines in $\mathbf{C}$.

If all λ_k are real and positive, then we note that $\mathbf{U}_L$ is simply $\mathbf{C}$ minus the negative real axis $\mathbf{R}_{\leq 0}$.

We also suppose given a sequence $A = \{a_k\}$ of distinct complex numbers. With L and A, we form the **asymptotic exponential polynomials** for integers $N \geq 1$:

$$Q_N(t) = a_0 + \sum_{k=1}^{N-1} a_k e^{-\lambda_k t}.$$

Throughout we shall also write

$$a_k = a(\lambda_k).$$

Similarly, we are given a sequence of complex numbers

$$\{p\} = \{p_0, \ldots, p_j, \ldots\}$$

with

$$\mathrm{Re}(p_0) \leq \mathrm{Re}(p_1) \leq \cdots \leq \mathrm{Re}(p_j) \leq \ldots$$

increasing to infinity. To every p in this sequence, we associate a polynomial B_p and, as before, we set

$$b_p(t) = B_p(\log t).$$

We then define the **asymptotic polynomials at** 0 to be

$$P_q(t) = \sum_{\mathrm{Re}(p)<\mathrm{Re}(q)} b_p(t)t^p.$$

In many, perhaps most, applications the exponents p are real. Because there are significant cases when the exponents are not necessarily real, we lay the foundations in appropriate generality. We define

$$m(q) = \max \deg B_p \quad \text{for} \quad \mathrm{Re}(p) = \mathrm{Re}(q).$$

Let $\mathbf{C}\langle T\rangle$ be the algebra of polynomials in T^p with arbitrary complex powers $p \in \mathbf{C}$. Then, with this notation,

$$P_q(t) \in \mathbf{C}[\log t]\langle t\rangle.$$

We let f be a piecewise continuous function on $(0,\infty)$ satisfying the following **asymptotic conditions** at infinity and zero.

AS 1. Given a positive number C and $t_0 > 0$, there exists N and $K > 0$ such that

$$|f(t) - Q_N(t)| \leq Ke^{-Ct} \text{ for } t \geq t_0.$$

AS 2. For every q, we have

$$f(t) - P_q(t) = O_q(t^{\mathrm{Re}(q)}|\log t|^{m(q)}) \text{ for } t \to 0,$$

where, as indicated, the implied constant depends on q.

Often we will write **AS 2** as

$$f(t) \sim \sum_p b_p(t)t^p.$$

Also, we will write $P_q(t) = P_q f(t)$ to denote the dependence on f. The crucial condition is **AS 2** and, in practice, is the most difficult to verify.

Let p_0 be an exponent of the asymptotic expansion **AS 2** with smallest (negative) value of $\mathrm{Re}(p_0)$. Let M be the largest integer $< -\mathrm{Re}(p_0)$. We call M the **reduced order** of the sequence $\{\lambda_k\}$. For instance, the sequence $\mathbf{Z}_{\geq 0}$ has reduced order 0.

The case when all a_k are non-negative integers will be called the **spectral case**. In such a situation one can view the coefficients a_k as determining a multiplicity in which the element λ_k appears in the sequence L.

The case when all the polynomials B_p are either 0 or constants, denoted by b_p, will be called the **special case**. In such a situation P_q is a polynomial with complex exponents and without the log terms.

Define $n(q)$ to be

$$n(q) = \max_{\mathrm{Re}(p)<\mathrm{Re}(q)} \deg B_p.$$

Throughout we will use p' to denote an element in the sequence $\{p\}$ with next largest real part for which $B_{p'}$ is not zero. In particular, this means that

$$n(q') = \max_{\mathrm{Re}(p)\leq\mathrm{Re}(q)} \deg B_p.$$

Theorem 1.5. *Let f satisfy* **AS 1** *and* **AS 2**. *Then* $\mathbf{LM}f$ *has a meromorphic continuation for $s \in \mathbf{C}$ and $z \in \mathbf{U}_L$. For each z, the function $s \mapsto \mathbf{LM}f(s,z)$ has poles only at the points $-(p+n)$ with $b_p \neq 0$ in the asymptotic expansion of f at 0. A pole at $-(p+n)$ has order at most $n(p)+1$. In the special case when the asymptotic expansion at 0 has no log terms, the poles are simple.*

Proof. We first do the analytic continuation in z, for $\mathrm{Re}(s)$ large. We subtract an exponential polynomial Q_N from $f(t)$, using the asymptotic axiom **AS 1**, to write

$$\begin{aligned}
\int_0^\infty f(t)e^{-zt}t^s\frac{dt}{t} &= \int_0^\infty [f(t)-Q_N(t)]e^{-zt}t^s\frac{dt}{t} + \int_0^\infty Q_N(t)c^{-zt}t^s\frac{dt}{t} \\
&= \int_0^\infty [f(t)-Q_N(t)]e^{-zt}t^s\frac{dt}{t} \\
&\quad + a_0\int_0^\infty e^{-zt}t^s\frac{dt}{t} + \sum_{k=1}^{N-1} a_k \int_0^\infty e^{-\lambda_k t}e^{-zt}t^s\frac{dt}{t} \\
&= \int_0^\infty [f(t)-Q_N(t)]e^{-zt}t^s\frac{dt}{t} \\
&\quad + a_0\frac{\Gamma(s)}{z^s} + \sum_{k=1}^{N-1} a_k\frac{\Gamma(s+p)}{(z+\lambda_k)^{s+p}}.
\end{aligned}$$

It is immediate that the terms involving the gamma function have the above stated meromorphy properties. In particular, note that the appearance of the term

$$\sum_{k=1}^{N-1} a_k \frac{\Gamma(s+p)}{(z+\lambda_k)^{s+p}}$$

requires us to restrict z to $\mathbf{U}_L$. So, at this time, it remains to study the integral involving $f - Q_N$. Since f satisfies **AS 2**, then so does $f - Q_N$, and the integral

$$\int_0^\infty [f(t) - Q_N(t)]e^{-zt}t^s \frac{dt}{t}$$

is absolutely convergent for $\mathrm{Re}(s) > -\mathrm{Re}(p_0)$ and $\mathrm{Re}(z) > -C$ if

$$f(t) - Q_N(t) = O(e^{-Ct}) \quad \text{for } t \to \infty.$$

By taking N sufficiently large, one can make C arbitrarily large, so this process shows how to meromorphically continue $\mathbf{LM}f(s,z)$ as a function of z. Next we show how to continue meromorphically in s.

For this we can apply Lemma 1.4 to $f - Q_N$, which shows that for $\mathrm{Re}(z) > -C$ the function

$$s \mapsto \mathbf{LM}f(s,z)$$

is meromorphic in $\mathbf{C}$. In addition, Lemma 1.4 shows that the only possible poles are as described in the statement of the theorem. Note that the set of these poles is discrete, because the values $p+n$ tend to infinity. This completes the proof of the theorem. □

We shall use a systematic notation for the coefficients of the Laurent expansion of $\mathbf{LM}f(s,z)$ near $s = s_0$. Namely we let $R_j(s_0; z)$ be the coefficient of $(s - s_0)^j$, so that

$$\mathbf{LM}_f(s,z) = \sum R_j(s_0; z)(s - s_0)^j.$$

We shall be particularly interested when $s_0 = 0$ or $s_0 = 1$. Also, when necessary, we will express the dependence of the coefficients on the function f by writing

$$R_{j,f}(s_0; z) = R_j(s_0; z).$$

Theorem 1.6. *Let f satisfy* **AS 1** *and* **AS 2**. *Then for every $z \in \mathbf{U}_L$ and s near 0, the function $\mathbf{LM}f(s,z)$ has a pole at $s=0$ of order at most $n(0')+1$, and the function $\mathbf{LM}f(s,z)$ has the Laurent expansion*

$$\mathbf{LM}f(s,z) = \frac{R_{-n(0')-1}(0;z)}{s^{n(0')+1}} + \cdots + R_0(0;z) + R_1(0;z)s + \ldots$$

where, for each $j<0$, $R_j(0;z) \in \mathbf{C}$ is a polynomial of degree $\leq -\mathrm{Re}(p_0)$.

Proof. For any $C>0$, choose N as in **AS 1** so that $f - Q_N$ is bounded as $t \to \infty$. Since

$$\begin{aligned}\mathbf{LM}f(s,z) &= \mathbf{LM}[f-Q_N](s,z) + \mathbf{LM}Q_N(s,z)\\ &= \mathbf{LM}[f-Q_N](s,z) + a_0\frac{\Gamma(s)}{z^s} + \sum_{k=1}^{N-1} a_k \frac{\Gamma(s+p)}{(z+\lambda_k)^{s+p}},\end{aligned}$$

it suffices to prove the theorem for $f - Q_N$, or, equivalently, assume that $f(t)$ is bounded as $t \to \infty$.

Let

$$P_{0'}(t) = \sum_{\mathrm{Re}(p)\leq 0} b_p(t)t^p,$$

so we include $\mathrm{Re}(p)=0$ in the sum. We decompose the integral into a sum:

$$\begin{aligned}\mathbf{LM}f(s,z) &= \int_0^\infty f(t)e^{-zt}t^s\frac{dt}{t}\\ &= \int_0^\infty [f(t)-P_{0'}(t)]e^{-zt}t^s\frac{dt}{t}\\ &\quad + \int_1^\infty f(t)e^{-zt}t^s\frac{dt}{t} + \int_0^1 P_{0'}(t)e^{-zt}t^s\frac{dt}{t}.\end{aligned}$$

The first two integrals are holomorphic at $s=0$ and $\mathrm{Re}(z)>0$.

By Lemma 1.2, the third integral is simply

$$\sum_{\mathrm{Re}(p)\leq 0} B_p(\partial_s)\int_0^1 e^{-zt}t^{s+p}\frac{dt}{t}$$

$$= \sum_{\mathrm{Re}(p)\leq 0}\sum_{k=0}^{\infty}\frac{(-z)^k}{k!}B_p(\partial_s)\left[\frac{1}{s+p+k}\right]$$

$$= \sum_{\mathrm{Re}(p)+k=0}\frac{(-z)^k}{k!}B_p(\partial_s)\left[\frac{1}{s}\right] + h_0(z) + O(s), \tag{4}$$

where $h_0(z)$ is entire in z. This proves the theorem, and (4) gives us an explicit determination of the polynomials $R_j(0;z)$ for $j<0$. □

The constant term $R_0(s_0;z)$ is so important that we give it a special notation; namely, for a meromorphic function $G(s)$ we let

$$\mathrm{CT}_{s=s_0}G(s) = \text{constant term in the Laurent expansion of } G(s) \text{ at } s=s_0.$$

That is,

$$\mathrm{CT}_{s=0}\mathbf{LM}f(s,z) = R_0(0;z).$$

Corollary 1.7. *Define*

$$\zeta_f(s,z) = \frac{1}{\Gamma(s)}\mathbf{LM}f(s,z) \quad \textit{and} \quad \xi_f(s,z) = \mathbf{LM}f(s,z).$$

Then, in the special case, $\zeta_f(s,z)$ is holomorphic at $s=0$ for $z\in\mathbf{U}_L$ and

$$\zeta_f'(0,z) = \mathrm{CT}_{s=0}\xi_f(s;z) + \gamma R_{-1,f}(0;z).$$

Proof. In the special case $n(0')=0$, so $\mathbf{LM}f(s,z)$ has a pole at $s=0$ of order at most 1. Since

$$\frac{1}{\Gamma(s)} = s+\gamma s^2 + O(s^3),$$

we have

$$\zeta_f(s,z) = \frac{1}{\Gamma(s)}\xi_f(s,z)$$
$$= R_{-1,f}(0;z) + (\mathrm{CT}_{s=0}\xi_f(s,z) + \gamma R_{-1,f}(0;z))s + O(s^2),$$

from which the corollary follows. □

Theorem 1.8. *The function* $\mathbf{LM}f$ *satisfies the equation*

$$\partial_z \mathbf{LM}f(s,z) = -\mathbf{LM}f(s+1,z), \quad \text{for } s \in \mathbf{C} \text{ and } z \in \mathbf{U}_L.$$

Proof. This is immediate by differentiating under the integral sign for $\mathrm{Re}(s)$ and $\mathrm{Re}(z)$ sufficiently large, and follows otherwise by analytic continuation. □

Corollary 1.9. *For any integer* j, *we have*

$$\partial_z R_j(s_0; z) = -R_j(s_0+1, z).$$

In particular, for the constant terms, we have

$$\partial_z R_0(0; z) = -R_0(1, z).$$

Corollary 1.10. *The Mellin transform*

$$\xi_f(s) = \mathbf{M}f(s),$$

has a meromorphic continuation to $s \in \mathbf{C}$ *whose only possible poles are at* $s = -p$ *such that* $b_p \neq 0$.

Proof. Write $f = f_0 + f_1$ with $f_0 = Q_N$ and $f_1 = f - Q_N$ with N sufficiently large so that we can apply **AS 1** to f_1 with $C > 0$. Then $\mathbf{M}f_0$ is entire in s. As for f_1, we have, for any q,

$$\mathbf{M}f_1(s) = \int_0^1 [f_1(t) - P_q(t)]\, t^s \frac{dt}{t}$$

$$+ \sum_{\mathrm{Re}(p)<\mathrm{Re}(q)} B_p(\partial_s)\left[\frac{1}{s+p}\right] + \int_1^\infty f_1(t) t^s \frac{dt}{t}. \tag{5}$$

The first integral in (5) is holomorphic for $\mathrm{Re}(s) > -\mathrm{Re}(q)$, by Lemma 1.4. By the construction of f_1, the second integral in (5) is entire in s. The sum in (6) is meromorphic for all $s \in \mathbf{C}$ with possible poles at $s = -p$. With all this, the proof is complete. □

Given the sequences A and L as defined in **AS 1**, one can consider:

a Dirichlet series

$$\zeta_{L,A}(s) = \zeta(s) = \sum_{k=1}^{\infty} a_k \lambda_k^{-s},$$

a theta series

$$\theta_{L,A}(t) = \theta(t) = a_0 + \sum_{k=1}^{\infty} a_k e^{-\lambda_k t},$$

a reduced theta series

$$\theta_{L,A}^{(1)}(t) = \theta^{(1)}(t) = \sum_{k=1}^{\infty} a_k e^{-\lambda_k t},$$

and, more generally, for each positive integer N,

a truncated theta series

$$\theta_{L,A}^{(N)}(t) = \theta^{(N)}(t) = \sum_{k=N}^{\infty} a_k e^{-\lambda_k t}.$$

We shall assume throughout that the theta series converges absolutely for $t > 0$. From **DIR 1** it follows that the convergence of the theta series is uniform for $t \geq \delta > 0$ for every δ. We shall apply the above results to $f = \theta$ with the associated sequence of exponential polynomials $Q_N\theta$ being the natural ones, namely

$$Q_N\theta(t) = a_0 + \sum_{k=1}^{N-1} a_k e^{-\lambda_k t}.$$

Note that the above notation leads to the formula

$$\theta(t) - Q_N\theta(t) = \theta^{(N)}(t)$$

The absolute convergence of the theta series $\theta(t)$ describes a type of convergence of $\theta^{(N)}$ near infinity that is uniform for all N. The following condition describes a type of uniformity of the asymptotics of $\theta(t)$ near $t = 0$.

AS 3. Given $\delta > 0$, there exists an $\alpha > 0$ and a constant $C > 0$ such that for all N and $0 < t \leq \delta$ we have

$$\left|\theta^{(N)}(t)\right| = |\theta(t) - Q_N(t)| \leq C/t^{\alpha}.$$

We shall see that the three conditions **AS 1**, **AS 2** and **AS 3** on the theta series correspond to conditions on the zeta function describing the growth of the sequence $\{\lambda_k\}$ and also the sequence $\{a_k\}$. The condition **DIR 1** simply states that the sequence $\{\lambda_k\}$ converges to infinity, in some weak sense. The following condition gives a slightly stronger convergence requirement.

DIR 2.

(a) The Dirichlet series

$$\sum_k \frac{a_k}{\lambda_k^{\sigma}}$$

converges absolutely for some real σ. Equivalently, we can say that there exists some $\sigma_0 \in \mathbf{R}_{>0}$ such that

$$|a_k| = O(|\lambda_k|^{\sigma_0}) \text{ for } k \to \infty.$$

(b) The Dirichlet series

$$\sum_k \frac{1}{\lambda_k^{\sigma}}$$

converges absolutely for some real σ. Specifically, let σ_1 be a real number for which

$$\sum_k \frac{1}{|\lambda_k|^{\sigma_1}} < \infty.$$

Theorem 1.11. *Assume that the theta series*

$$\theta(t) = \sum_{k=1}^{\infty} a_k e^{-\lambda_k t}$$

satisfies **AS 1**, **AS 2** *and* **AS 3**, *and assume that* $\mathrm{Re}(\lambda_k) > 0$ *for all* $k > 0$. *Then for* $\mathrm{Re}(s) > \alpha$,

$$\xi(s) = \mathbf{M}\theta(s) = \Gamma(s) \sum_{k=1}^{\infty} \frac{a_k}{\lambda_k^s}$$

in the sense that the series on the right converges absolutely, to $\mathbf{M}\theta(s)$. *In particular, the convergence condition* **DIR 2(a)** *is satisfied for the Dirichlet series*

$$\sum_{k=1}^{\infty} \frac{a_k}{\lambda_k^s}.$$

Proof. For $\sigma = \mathrm{Re}(s) > \alpha$, we have

$$\xi(s) - \Gamma(s) \sum_{k=1}^{N-1} \frac{a_k}{\lambda_k^s} = \int_0^1 [\theta(t) - Q_N\theta(t)] t^s \frac{dt}{t} + \int_1^{\infty} [\theta(t) - Q_N\theta(t)] t^s \frac{dt}{t}.$$

From **AS 1** we have

$$\left| \int_1^{\infty} [\theta(t) - Q_N\theta(t)] t^s \frac{dt}{t} \right| \leq K \int_1^{\infty} e^{-Ct} t^{\sigma} \frac{dt}{t},$$

which goes to zero as $C \to \infty$. Similarly, by **AS 3** we have that

$$\left| \int_0^1 [\theta(t) - Q_N\theta(t)] t^s \frac{dt}{t} \right| \leq C \int_0^1 t^{\sigma - \alpha} \frac{dt}{t},$$

which is uniformly bounded if $\sigma > \alpha$. Therefore, we have that for $\sigma \geq \alpha + \epsilon > \alpha$, the difference

$$\xi(s) - \Gamma(s) \sum_{k=1}^{N-1} \frac{a_k}{\lambda_k^s}$$

is uniformly bounded for all N. By letting N approach ∞, we can interchange limit and integral, by dominated convergence, to show that for $\sigma \geq \alpha + \epsilon > \alpha$,

$$\lim_{N\to\infty}\left[\xi(s) - \Gamma(s)\sum_{k=1}^{N-1}\frac{a_k}{\lambda_k^s}\right] = 0,$$

which completes the proof of the theorem. □

Remark 1. If $-\mathrm{Re}(p_0)$ is an integer, then $M+2$ is the smallest integer m for which the Dirichlet series

$$\sum_1^\infty \frac{|a_k|}{|\lambda_k|^m} < \infty$$

converges. If $-\mathrm{Re}(p_0)$ is not an integer, then $M+1$ is the smallest such m. In any event, we let m_0 be the smallest such m. The exponent $\mathrm{Re}(p_0)$, which comes from **AS 2**, is not independent of the integer m_0, which comes from **DIR 2**. In fact

$$m_0 - 1 \leq -\mathrm{Re}(p_0) < m_0. \tag{6}$$

Indeed

$$\begin{aligned}\xi(s) &= \Gamma(s)\sum_{k=1}^\infty a_k\lambda_k^{-s} \\ &= \sum_{\mathrm{Re}(p)<\mathrm{Re}(q)}\int_0^1 b_p(t)t^{p+s}\frac{dt}{t} + \int_1^\infty [\theta(t) - P_q(t)]t^s\frac{dt}{t}.\end{aligned}$$

The second integral on the right is holomorphic for

$$\mathrm{Re}(s) > -\mathrm{Re}(q).$$

The first integral on the right has its first pole at $s + p_0 = 0$, so at $-p_0$, whence $m_0 > -\mathrm{Re}(p_0)$. The first inequality in (6) follows from the minimality of m_0 since $\mathrm{Re}(p_0) \leq 0$.

The following condition on the sequence L requires that, beyond what is stated in the convergence condition **DIR 1**, the sequence λ_k approaches infinity in a sector.

DIR 3. There is a fixed $\epsilon > 0$ such that for all k sufficiently large, we have

$$-\frac{\pi}{2} + \epsilon \leq \arg(\lambda_k) \leq \frac{\pi}{2} - \epsilon.$$

Equivalently, there exists positive constants C_1 and C_2 such that for all k sufficiently large,

$$C_1|\lambda_k| \leq \mathrm{Re}(\lambda_k) \leq C_2|\lambda_k|.$$

Theorem 1.12. *Let (L, A) be sequences for which the associated Dirichlet series*

$$\sum_{k=1}^{\infty} \frac{a_k}{\lambda_k^s}$$

satisfies the three convergence conditions **DIR 1**, **DIR 2** *and* **DIR 3**. *Then the theta series*

$$\theta(t) = \sum_{k=1}^{\infty} a_k e^{-\lambda_k t}$$

satisfies **AS 1** *and* **AS 3**.

Proof. Let us first show how **AS 3** follows. Directly from **DIR 2** and **DIR 3** we have, for some constants c_1 and c_2, the inequalities

$$\begin{aligned}\left|\theta(t) - Q_N\theta(t)\right| &\leq \left|\sum_{k=N}^{\infty} a_k e^{-\lambda_k t}\right| \\ &\leq \sum_{k=N}^{\infty} |a_k| e^{-\mathrm{Re}(\lambda_k)t} \\ &\leq c_1 \sum_{k=N}^{\infty} |\lambda_k|^{\sigma_0} e^{-c_2|\lambda_k|t}.\end{aligned}$$

Note that for any $x \geq 0$, there is a constant $c = c(\sigma_0 + \sigma_1)$ such that

$$x^{\sigma_0+\sigma_1} e^{-x} \leq c.$$

Let us apply this inequality to $x = c_2|\lambda_k|t$ and then sum for $k \geq N$ to obtain, for any $t > 0$,

$$\begin{aligned} \left|\theta(t) - Q_N\theta(t)\right| &\leq c_1 \sum_{k=N}^{\infty} |\lambda_k|^{\sigma_0} e^{-c_2|\lambda_k|t} \\ &\leq c_1 \cdot c(c_2 t)^{-\sigma_0-\sigma_1} \sum_{k=N}^{\infty} |\lambda_k|^{-\sigma_1}. \end{aligned} \tag{7}$$

Now if we let

$$\alpha = \sigma_0 + \sigma_1,$$

and

$$C = c_1 c(c_2)^{-\alpha} \sum_{k=1}^{\infty} |\lambda_k|^{-\sigma_1},$$

then (7) becomes

$$\left|\theta(t) - Q_N\theta(t)\right| \leq C/t^{\alpha},$$

which establishes the asymptotic condition **AS 3**.

In order to establish **AS 1**, we need some preliminary calculations. First, note that by choosing $c_3 < c_2$, we can write

$$\left|\theta(t) - Q_N\theta(t)\right| \leq \left|\sum_{k=N}^{\infty} a_k e^{-\lambda_k t}\right| \leq c_1' \sum_{k=N}^{\infty} e^{-c_3|\lambda_k|t}.$$

Now let

$$C_N = \min\{c_3|\lambda_k|\} \quad \text{for} \quad k \geq N,$$

and

$$t_0^{(N)} = \max\left\{\frac{\log|\lambda_k|\sigma_1}{c_3|\lambda_k| - \frac{1}{2}C_N}\right\} \quad \text{for} \quad k \geq N. \tag{8}$$

By **DIR 1** and **DIR 3** we have

$$\lim_{N\to\infty} C_N = \infty,$$

and since

$$\frac{1}{2}C_N \leq \frac{1}{2}c_3|\lambda_k| \quad \text{for} \quad k \geq N,$$

we can write

$$c_3|\lambda_k| - \frac{1}{2}C_N \geq c_3|\lambda_k| - \frac{1}{2}c_3|\lambda_k| = \frac{1}{2}c_3|\lambda_k| > 0.$$

Therefore,

$$\frac{\log|\lambda_k|\sigma_1}{c_3|\lambda_k| - \frac{1}{2}C_N} \leq \frac{\log|\lambda_k|}{|\lambda_k|} \cdot \frac{2\sigma_1}{c_3} \quad \text{for} \quad k \geq N. \tag{9}$$

By combining (8) and (9) we conclude that

$$t_0^{(N)} \leq \max\left\{\frac{\log|\lambda_k|}{|\lambda_k|}\right\} \cdot \frac{2\sigma_1}{c_3} \quad \text{for} \quad k \geq N,$$

so, in particular,

$$\lim_{N\to\infty} t_0^{(N)} = 0.$$

Therefore, there exists $t_0 < \infty$ such that

$$t_0^{(N)} \leq t_0 \quad \text{for all } N.$$

Note that for $t > t_0$ and any $k \geq N$ we have

$$\begin{aligned}
(c_3|\lambda_k| - \frac{1}{2}C_N)t &\geq (c_3|\lambda_k| - \frac{1}{2}C_N)t_0 \\
&\geq (c_3|\lambda_k| - \frac{1}{2}C_N)t_0^{(N)} \\
&\geq \frac{c_3|\lambda_k| - \frac{1}{2}C_N}{c_3|\lambda_k| - \frac{1}{2}C_N}\log|\lambda_k|\sigma_1 \\
&\geq \log|\lambda_k|\sigma_1,
\end{aligned}$$

Therefore, for $t > t_0$,

$$e^{-(c_3|\lambda_k| - \frac{1}{2}C_N)t} \leq |\lambda_k|^{-\sigma_1},$$

which gives the bound

$$\left|\theta(t) - Q_N\theta(t)\right| \le c_1' e^{-\frac{1}{2}C_N t} \sum_{k=N}^{\infty} e^{-(c_3|\lambda_k| - \frac{1}{2}C_N)t}$$
$$\le c_1' e^{-\frac{1}{2}C_N t} \sum_{k=N}^{\infty} |\lambda_k|^{-\sigma_1},$$

which shows that **AS 1** holds, and completes the proof of the theorem. □

Remark 2. The convergence condition **DIR 1** is assumed as part of the asymptotic condition **AS 1**. Theorem 1.11 asserts that if the theta series

$$\theta(t) = \sum_{k=1}^{\infty} a_k e^{-\lambda_k t}$$

satisfies the asymptotic conditions **AS 1**, **AS 2** and **AS 3**, then

$$\xi(s) = \mathbf{M}\theta(s) = \Gamma(s) \sum_{k=1}^{\infty} \frac{a_k}{\lambda_k^s}.$$

Further, the sequences (L, A) satisfies the convergence condition **DIR 2(a)** and, by Corollary 1.10, $\xi(s)$ has a meromorphic continuation to all $s \in \mathbf{C}$. Theorem 1.12 states that if the two sequences (L, A) satisfy the three convergence conditions **DIR 1**, **DIR 2** and **DIR 3**, then the corresponding theta series satisfies the asymptotic conditions **AS 1** and **AS 3**. As previously stated, the condition **AS 2** is quite delicate and, in practice, is the most difficult to verify. In §7 we will show, under additional meromorphy and growth condition hypothesis on ξ, **AS 2** follows from the three convergence conditions **DIR 1**, **DIR 2** and **DIR 3**.

§2. Laurent expansion at $s = 0$, Weierstrass product, and the Lerch formula

In this section, we consider the case when a_k is a non-negative integer for all k, which we define to be **the spectral case.** The corresponding Dirichlet series is then also called spectral, or **the spectral zeta function.** We develop some ideas of Voros [Vo 87], especially about his formula (4.1) which we formulate in a general context below as Theorem 2.1. Our arguments are somewhat different from those of Voros, who makes "no pretence of rigour". Basically, we want to make sense out of an infinite product of a sequence of complex numbers $L = \{\lambda_k\}$, counted with multiplicities $A = \{a_k\}$, which satisfies certain conditions. The results in §1 and the above definitions establish a line of investigation via a zeta function. In this section we indicate the line along Weierstrass products, and we show how the two approaches connect.

We recall the construction of a Weierstrass product. Let λ be a non-zero complex number, let m be an integer $\geq m_0$, and let

$$E_m(z, \lambda) = \left(1 - \frac{z}{\lambda}\right) \exp\left(\frac{z}{\lambda} + \frac{1}{2}\left(\frac{z}{\lambda}\right)^2 + \ldots + \frac{1}{m-1}\left(\frac{z}{\lambda}\right)^{m-1}\right),$$

or also

$$\log E_m(z, \lambda) = -\sum_{n=m}^{\infty} \frac{1}{n}\left(\frac{z}{\lambda}\right)^n.$$

We shall work under the assumptions **DIR 2** and **DIR 3**, and, for $m \geq m_0$, we define the **Weierstrass product**

$$D_{m,L}(z) = z^{a_0} \prod_{k=1}^{\infty} E_m(z, -\lambda_k)^{a_k}.$$

By the elementary theory of Weierstrass products, $D_{m,L}(z)$ is an entire function of strict order $\leq m$, that is

$$\log |D_{m,L}(z)| = O(|z|^m) \quad \text{for} \quad |z| \to \infty.$$

We use the adjective "strict" to avoid putting an ε in the exponent on the right hand side.

Let $D(z)$ be an entire function of strict order $\leq m$ with the same zeros as $D_{m,L}(z)$, counting multiplicities. Then there exists a polynomial $P_D(z)$ of degree $\leq m$ such that

$$D(z) = e^{P_D(z)} D_{m,L}(z).$$

We shall describe conditions that determine the polynomial P_D uniquely.

Suppose the three convergence conditions **DIR 1**, **DIR 2** and **DIR 3** are satisfied, so we have the spectral zeta function and the numbers

$$\zeta(n) = \sum_{k=1}^{\infty} \frac{a_k}{\lambda_k^n} \quad \text{for} \quad n \geq m.$$

We then have the power series of $D(z)$ at the origin coming from the expansion

$$\begin{aligned} (1) \qquad \log D(z) &= a_0 \log z + P_D(z) + \sum_{n=m}^{\infty} (-1)^{n-1} \zeta(n) \frac{z^n}{n} \\ &= a_0 \log z + c_0 + \sum_{n=1}^{\infty} (-1)^{n-1} c_n \frac{z^n}{n} \end{aligned}$$

where the coefficients c_n satisfy

$$c_n = \zeta(n) \text{ for } n \geq m+1.$$

If $D(0) \neq 0$, meaning $a_0 = 0$, then $D(0) = e^{c_0}$, and we have

$$(2) \qquad -\log [D(z)/D(0)] = \sum_{n=1}^{\infty} (-1)^n c_n \frac{z^n}{n}.$$

From this we see that the coefficients of $P_D(z)$ can be determined from the expansion (2) and the numbers

$$\zeta(m_0), \ldots, \zeta(m).$$

Since $D_{m_0,L}(z)$ is an entire function, the logarithmic derivative

$$\frac{d}{dz} \log D_{m_0,L}(z) = D'_{m_0,L}/D_{m_0,L}(z)$$

is meromorphic, and further derivatives $(d/dz)^r \log D_{m_0,L}(z)$ are also meromorphic, immediately expressible as sums that are of Mittag-Leffler type as follows. For an integer $r \geq 1$ let us define

$$T_{r,L}(z) = T_r(z) = \frac{(-1)^{r-1}}{\Gamma(r)} \left(\frac{d}{dz}\right)^r \log D_{m_0,L}(z).$$

We have trivially

$$\frac{d}{dz} T_r(z) = -rT_{r+1}(z).$$

Since the logarithmic derivative transforms products to sums, as part of the standard elementary theory of Weierstrass products we obtain for each $r \geq 1$ the expansion:

(3)

$$T_r(z) = \begin{cases} \dfrac{a_0}{z^r} + \displaystyle\sum_{k=1}^{\infty} \left[\frac{a_k}{(z+\lambda_k)^r} - \sum_{n=0}^{N} \binom{-r}{n} \frac{a_k z^n}{\lambda_k^{n+r}} \right]; & r < m_0 \\ \dfrac{a_0}{z^r} + \displaystyle\sum_{k=1}^{\infty} \frac{a_k}{(z+\lambda_k)^r}; & r \geq m_0 \end{cases}$$

where $N = m_0 - r - 1$. In (3) we have the usual binomial coefficient

$$\binom{-r}{n} = \frac{\Pi_n(r)}{n!}$$

where

$$\Pi_n(r) = (-1)^n r(r+1)\dots(r+n-1).$$

Note that for $n > 1$ we have

$$\Pi_n(r) = -r\Pi_{n-1}(r+1).$$

Let us define $\Pi_n(r)$ for negative n through this recursive relation.

Because of the absolute convergence of the theta series $\theta(t)$, we have, by Theorem 1.8, for all z such that $\mathrm{Re}(z + \lambda_k) > 0$ for all k, and any $r \geq m_0$,

$$T_r(z) = \frac{a_0}{z^r} + \sum_{k=1}^{\infty} \frac{a_k}{(z+\lambda_k)^r} = \frac{1}{\Gamma(r)} \int_0^{\infty} \theta(t) e^{-zt} t^r \frac{dt}{t}.$$

If we let

$$\theta_z(t) = e^{-zt}\theta(t) = a_0 e^{-zt} + \sum_{k=1}^{\infty} a_k e^{-(z+\lambda_k)t},$$

we can write $T_r(z)$ as a Mellin transform:

$$T_r(z) = \frac{1}{\Gamma(r)} \int_0^{\infty} \theta_z(t) t^r \frac{dt}{t}.$$

The convergence of the integral and sum is uniform for $\mathrm{Re}(z)$ sufficiently large.

We shall apply Theorem 1.8 with $f = \theta$ and $\mathbf{LM}f = \xi$ to obtain the following theorem, which we call the **Lerch formula**.

Theorem 2.1. *In the spectral case, assume that the theta function θ satisfies the asymptotic conditions* **AS 1**, **AS 2** *and* **AS 3** *with the natural sequence of exponential polynomials (4). Then there exists a unique polynomial $P_L(z)$ of degree $\leq m_0 - 1$ such that if we define*

$$D_L(z) = e^{P_L(z)} D_{m_0,L}(z),$$

then for all $z \in \mathbf{C}$ with $\mathrm{Re}(z)$ sufficiently large, we have

$$D_L(z) = \exp(-\mathrm{CT}_{s=0}\xi(s,z)).$$

Hence $\exp(-\mathrm{CT}_{s=0}\xi(s,z))$ has an analytic continuation to all $z \in \mathbf{C}$ to the entire function $D_L(z)$. In particular,

$$D'_L/D_L(z) = -\partial_z \mathrm{CT}_{s=0}\xi(s,z) = \mathrm{CT}_{s=1}\xi(s,z) = R_0(1;z).$$

Proof. If we count $\lambda_0 = 0$ with multiplicity a_0, we find for $\mathrm{Re}(s)$ large and $r \geq m_0$,

$$(\partial_z)^r \xi(s,z) = \Gamma(s) \sum_{k=0}^{\infty} (-s)(-s-1)\dots(-s-r+1) \frac{a_k}{(z+\lambda_k)^{s+r}}$$

$$= \Gamma(s)\Pi_r(s) \sum_{k=0}^{\infty} \frac{a_k}{(z+\lambda_k)^{s+r}}$$

where, as before,

$$\Pi_r(s) = (-1)^r s(s+1)\dots(s+r-1).$$

We now look at the constant term in a Laurent expansion at $s = 0$. If $r > m_0$, then the series

$$\sum_{k=0}^{\infty} \frac{a_k}{(z+\lambda_k)^r}$$

converges. At $s = 0$ the gamma function has a first order pole with residue 1. Note that the $\Pi_r(0) = 0$. So, Theorem 1.6 gives us a bound $m_0 - 1$ for the degrees of the polynomials in z occuring as coefficients in the negative powers in s of the Laurent expansion of $\xi(s,z)$. From this, we conclude that $(\partial_z)^r \xi(s,z)$ is holomorphic near $s = 0$. With all this, we can set $s = 0$ and obtain the equality

$$(\partial_z)^r \mathrm{CT}_{s=0}\xi(s,z) = (-1)^r \Gamma(r) \sum_{k=0}^{\infty} \frac{a_k}{(z+\lambda_k)^r},$$

from which we obtain, using (3),

$$(\partial_z)^r \left[\log D_{m_0}(z) + \mathrm{CT}_{s=0}\xi(s,z)\right] = 0.$$

Hence, there is a polynomial $P_L(z)$ of degree $\leq r-1$ such that

$$\log D_{m_0,L}(z) + P_L(z) = -\mathrm{CT}_{s=0}\xi(s,z)$$

for $\mathrm{Re}(z)$ sufficiently large, which completes the proof of the theorem upon setting $r = m_0$. □

We call $D(z) = D_L(z)$ the **regularized product** associated to the sequence $L + z$. In particular, $D(0)$ is the regularized product of the sequence L.

Remark 1. Assume L can be written as the disjoint union of the sequences L' and L'' where L' satisfies the convergence conditions **DIR 1**, **DIR 2** and **DIR 3**, and $\theta_{L'}(t)$ satisfies the asymptotic conditions **AS 1**, **AS 2** and **AS 3**. Then L'' and $\theta_{L''}(t)$ necessarily satisfy these conditions and

$$\theta(t) = \theta_{L'}(t) + \theta_{L''}(t).$$

From this, we immediately have

$$\xi(s, z) = \xi_{L'}(s, z) + \xi_{L''}(s, z)$$

and

$$D_L(z) = D_{L'}(z) D_{L''}(z).$$

A particular example of such a decomposition is the case when L' is a finite subset of L, in which case $D_{L'}(z) = \prod[(z + \lambda_k)e^{\gamma}]$; see Remark 2 below.

It now becomes of interest to determine in some fashion the coefficients of the polynomial $P_L(z)$, and for this it suffices to determine $P_L^{(r)}(0)$. We define the **reduced sequence** L_0 to be the sequence which is the same as L except that we delete $\lambda_0 = 0$. Then

$$\theta_{L_0}(t) = \theta(t) - a_0$$

and, as discussed in Remark 1, we have

$$D_{m_0, L_0}(z) = z^{-a_0} D_{m_0, L}(z).$$

Theorem 2.2. *The polynomials P_L and P_{L_0}, as defined in Theorem 2.1, are equal. If we let ∂_2 be the partial derivative with respect to the second variable, then*

$$P_{L_0}^{(r)}(0) = -\partial_2^r \mathrm{CT}_{s=0}\xi(s, 0)$$

for $0 \leq r \leq m_0 - 1$. That is

$$P_{L_0}(z) = -\sum_{k=0}^{m_0-1} \partial_2^k \mathrm{CT}_{s=0}\xi(s, 0) \frac{z^k}{k!}.$$

Proof. From Theorem 2.1 let us write

$$P_{L_0}(z) = -\log D_{m_0, L_0}(z) - \mathrm{CT}_{s=0}\xi(s, 0).$$

The canonical product of $D_{m_0,L_0}(z)$ is such that the Taylor series of $\log D_{m_0,L_0}(z)$ at the origin begins with powers of z which are at least z^{m_0}. Hence,

$$\left(\frac{\partial}{\partial z}\right)^r \log D_{m_0,L_0}(z)\Bigg|_{z=0} = 0 \quad \text{for } 0 \leq r \leq m_0 - 1.$$

This proves the theorem. □

Remark 2. The normalization $D_L(z)$ that we have given is the most convenient one for the formalism we are developing. One may also define the **characteristic determinant** $\mathbf{D}_L(z)$ by the condition

$$-\log \mathbf{D}_L(z) = R_{1,\zeta}(0;z) = \mathrm{CT}_{s=0}\left[s^{-1}\zeta(s,z)\right] = \mathrm{CT}_{s=0}\left[\frac{\xi(s,z)}{s\Gamma(s)}\right],$$

so $R_{1,\zeta}(0;z)$ is the coefficient of s in the Laurent expansion of $\zeta(s,z)$ at $s=0$. In the special case we have that

$$R_{1,\zeta}(0;z) = \zeta'(0,z).$$

Note that $\log \mathbf{D}_L(z)$ and $\log D_L(z)$ differ by an obvious polynomial in z coming from the Laurent expansion of $\Gamma(s)$ at $s = 0$. For example, if L is a finite sequence, then $\mathbf{D}_L(z) = \prod(z + \lambda_k)$. Also, let us record the formula

$$\begin{aligned}\zeta'(0,z) &= R_{0,\xi}(0;z) + \gamma R_{-1,\xi}(0;z)\\ &= \mathrm{CT}_{s=0}\xi(s,z) + \gamma R_{-1,\xi}(0;z),\end{aligned}$$

which, again, holds only in the special case.

Example 1. Theorems 2.1 and 2.2 provide a general setting for the some classical formulas. Voros [Vo 87] gives examples, including the simplest case which concerns the sequence $\lambda_k = k$ and the gamma function (see his Example c). Theorems 2.1 and 2.2 contain as a special case the classical Lerch formula, which states that

$$\log \mathbf{D}(z) = -\zeta'_{\mathbf{Q}}(0,z) \quad \text{if} \quad \mathbf{D}(z) = \sqrt{2\pi}/\Gamma(z)$$

and

$$\xi_{\mathbf{Q}}(s,z) = \Gamma(s)\sum_{n=0}^{\infty}\frac{1}{(z+n)^s} = \Gamma(s)\zeta_{\mathbf{Q}}(s,z).$$

Indeed, we have the immediate relation

$$\log \mathbf{D}(z) = \log D(z) + (\gamma + 1)z - \gamma/2$$

coming from the definitions and the expansion

$$\Gamma(s) = \frac{1}{s} - \gamma + O(s)$$

at $s = 0$.

Example 2. Let $\zeta_{\mathbf{Q}}$ be the Riemann zeta function, and let $\{\rho_k\}$ be the sequence of zeros in the critical strip with $\mathrm{Im}(\rho_k) > 0$. Let $\lambda'_k = \rho_k/i$. It is a corollary of a theorem of Cramér [Cr 19] that the theta function

$$\theta(t) = \sum a_k e^{-\lambda'_k t}$$

satisfies **AS 2**. The Lerch formula for the sequence $\{\lambda'_k\}$ (with multiplicities a_k) then specializes to a formula discovered by Deninger [De 92], Theorem 3.3. In [JoL 92b] we extend Cramér's theorem to a wide class of functions having an Euler product and functional equation, including the L-series of a number field (Hecke and Artin), L-functions arising in representation theory and modular forms, Selberg-type zeta funtions and L-functions for Riemann surfaces and certain higher dimensional manifolds, etc. The present section therefore applies to these functions as well, and thus the Lerch formula is valid for them. Note that Deninger's method did not give him the analytic continuation for the expressions in his formula, but such continuation occurs naturally in our approach.

§3. Expressions at $s = 1$.

In this section we return to general Dirichlet series, as in §1, meaning we are given sequences (L, A) whose Dirichlet series that satisfy the convergence conditions **DIR 1**, **DIR 2** and **DIR 3**, with associated theta series $\theta_{L,A} = \theta$ that satisfies the three asymptotic conditions **AS 1**, **AS 2**, and **AS 3**. With this, we will study the xi-function

$$\xi(s, z) = \mathbf{LM}\theta(s, z) = \int_0^\infty \theta(t)e^{-zt}t^s \frac{dt}{t}$$

near $s = 1$. Referring to **AS 2**, let us define the **principal part of the theta function** $P_0\theta(t)$ by

$$P_0\theta(t) = \sum_{\mathrm{Re}(p)<0} b_p(t)t^p. \tag{1}$$

Recall that, by definition, $m(0) = \max \deg B_p$ for $\mathrm{Re}(p) = 0$ so

$$\theta(t) - P_0\theta(t) = \begin{cases} O(1) & \text{in the special case} \\ O(|\log t|^{m(0)}) & \text{in the general case,} \end{cases} \tag{2}$$

as t approaches zero.

Theorem 3.1. *Let C and N be related as in* **AS 1**. *Then the function*

$$\xi(s, z) - \Gamma(s)\sum_{k=0}^{N-1} \frac{a_k}{(\lambda_k + z)^s} - \int_0^1 P_0\theta(t)e^{-zt}t^s \frac{dt}{t}$$

has a holomorphic continuation to the region

$$\{\mathrm{Re}(s) > 0\} \times \{\mathrm{Re}(z) > -C\}.$$

Proof. Let us write

$$\xi(s,z) - \Gamma(s)\sum_{k=0}^{N-1}\frac{a_k}{(\lambda_k+z)^s} - \int_0^1 P_0\theta(t)e^{-zt}t^{s+p}\frac{dt}{t}$$

$$= \int_0^\infty [\theta(t) - Q_N\theta(t)]\, e^{-zt}t^s\frac{dt}{t} - \int_0^1 P_0\theta(t)e^{-zt}t^s\frac{dt}{t}$$

$$(3)\qquad = \int_0^1 [\theta(t) - P_0\theta(t)]\, e^{-zt}t^s\frac{dt}{t}$$

$$(4)\qquad + \int_1^\infty [\theta(t) - Q_N\theta(t)]\, e^{-zt}t^s\frac{dt}{t}$$

$$(5)\qquad - \int_0^1 Q_N\theta(t)e^{-zt}t^s\frac{dt}{t}.$$

Using **AS 2**, we have that the integral in (3) is holomorphic for all $z \in \mathbf{C}$ and $\mathrm{Re}(s) > 0$, as is the integral in (5), and the integral in (4) is holomorphic for all $s \in \mathbf{C}$ and $\mathrm{Re}(z) > -C$. Combining this, the stated claim has been proved. □

Corollary 3.2. *The function*

$$\xi(s,z) - \int_0^1 P_0\theta(t)e^{-zt}t^s\frac{dt}{t}$$

is meromorphic at $s = 1$ for all z with singularities that are simple poles at $z = -\lambda_k$. Also, the residue at $z = -\lambda_k$ is equal to a_k.

Proof. Immediate from the proof of Theorem 3.1 by taking C, consequently N, sufficiently large. □

We now study the **singular term**

$$\int_0^1 P_0\theta(t)e^{-zt}t^s\frac{dt}{t} = \sum_{\mathrm{Re}(p)<0}\int_0^1 b_p(t)e^{-zt}t^s\frac{dt}{t}$$

that appears in Corollary 3.2. To do so, we expand e^{-zt} in a power series and apply the following lemma.

Lemma 3.3. *Given p, there is an entire function $h_p(z)$ such that as s approaches 1, such that:*

(a) *Special Case:*

$$\int_0^1 b_p e^{-zt} t^{s+p} \frac{dt}{t} = \sum_{k=0}^{\infty} \frac{(-z)^k}{k!} b_p \cdot \frac{1}{s+p+k}$$

$$= \begin{cases} \dfrac{(-z)^{-p-1}}{(-p-1)!} b_p \cdot \dfrac{1}{s-1} + h_p(z) + O(s-1), & p \in \mathbf{Z}_{<0} \\ h_p(z) + O(s-1), & p \notin \mathbf{Z}_{<0}. \end{cases}$$

(b) *General Case:*

$$\int_0^1 b_p(t) e^{-zt} t^{s+p} \frac{dt}{t} = \sum_{k=0}^{\infty} \frac{(-z)^k}{k!} B_p(\partial_s) \left[\frac{1}{s+p+k}\right]$$

$$= \begin{cases} \dfrac{(-z)^{-p-1}}{(-p-1)!} B_p(\partial_s) \left[\dfrac{1}{s-1}\right] + h_p(z) + O(s-1), & p \in \mathbf{Z}_{<0} \\ h_p(z) + O(s-1), & p \notin \mathbf{Z}_{<0}. \end{cases}$$

From Lemma 3.3, we can assert the existence of an entire function $h(z)$ such that the singular term can be written as

$$\int_0^1 P_0\theta(t) e^{-zt} t^s \frac{dt}{t} = \sum_{p+k=-1} \frac{(-z)^k}{k!} B_p(\partial_s) \left[\frac{1}{s-1}\right] + h(z) + O(s-1).$$

The following theorem, which is the main result of this section, then follows directly from Theorem 3.1 and Lemma 3.3.

Theorem 3.4. *Near $s = 1$, the Hurwitz xi function $\xi(s,z)$ has the expansion*

$$\xi(s,z) = \frac{R_{-n(1)-1}(1;z)}{(s-1)^{n(1)+1}} + \cdots + \frac{R_{-1}(1;z)}{(s-1)} + R_0(1;z) + O(s-1),$$

where:

(a) *For $j < 0$, $R_j(1;z)$ is a polynomial of degree $< -\mathrm{Re}(p_0)$; in fact, the polar part of $\xi(s,z)$ near $s = 1$ is expressed by*

$$\frac{R_{-n(1)-1}(1;z)}{(s-1)^{n(1)+1}} + \cdots + \frac{R_{-1}(1;z)}{(s-1)} = \sum_{p+k=-1} \frac{(-z)^k}{k!} B_p(\partial_s)\left[\frac{1}{s-1}\right];$$

(b) *$R_0(1;z) = \mathrm{CT}_{s=1}\xi(s,z)$ is a meromorphic function in z for all $z \in \mathbf{C}$ whose singularities are simple poles at $z = -\lambda_k$ with residue equal to a_k. Furthermore,*

$$\mathrm{CT}_{s=1}\xi(s,z) = -\partial_z \mathrm{CT}_{s=0}\xi(s,z).$$

In the special case, the expansion of $\xi(s,z)$ near $s = 1$ simplifies to

$$\xi(s,z) = \frac{R_{-1}(1;z)}{s-1} + R_0(1;z) + O(s-1),$$

with

$$R_{-1}(1;z) = \sum_{p+k=-1} b_p \frac{(-z)^k}{k!}.$$

Let (L, A) be sequences of complex numbers that satisfy the three convergence conditions **DIR 1**, **DIR 2** and **DIR 3** and such that the associated theta function

$$\theta_{L,A}(t) = \theta(t) = \sum_{k=1}^{\infty} a_k e^{-\lambda_k t}$$

satisfies the three asymptotic conditions **AS 1**, **AS 2** and **AS 3**. We define the **regularized harmonic series** $R(z)$ associated to (L, A) to be

$$R_{L,A}(z) = R(z) = \mathrm{CT}_{s=1}\xi(s,z) = -\partial_z \mathrm{CT}_{s=0}\xi(s,z).$$

Remark 1. Theorem 3.4 states that the regularized harmonic series associated to (L, A) is a meromorphic function in z whose singularities are simple poles at $z = -\lambda_k$ and corresponding residues equal to a_k. Further, by **DIR 2** and **DIR 3**, Theorem 1.8 and Theorem 1.12, we have, for any integer $n \geq m_0$, the expression

$$\partial_z^n R(z) = (-1)^n \mathrm{CT}_{s=1} \xi(s+n, z) = (-1)^n \Gamma(n) \sum_{k=1}^{\infty} \frac{a_k}{(z+\lambda_k)^n}.$$

If all numbers a_k are integers, then one can assert the existence of a meromorphic function $D(z)$, unique up to constant factor, which satisfies the relation

$$D'/D(z) = R(z). \tag{6}$$

The Spectral Case. In this case, with $a_k \in \mathbf{Z}_{>0}$ for all k, Theorem 3.4 asserts the existence of an entire function $D(z)$, unique up to constant factor, such that

$$D'/D(z) = \mathrm{CT}_{s=1} \xi(s, z). \tag{7}$$

Further, we have, by the Lerch formula (Theorem 2.1),

The Basic Identity:

$$R(z) = \mathrm{CT}_{s=1} \xi_L(s, z) = -\partial_z \mathrm{CT}_{s=0} \xi_L(s, z) = D_L'/D_L(z).$$

We shall normalize the constant factor in (7) so that

$$D(z) = D_L(z).$$

Next we shall give another type of expression for the singular term, which also gives an expression for $\mathrm{CT}_{s=1} \xi_L(s, z)$ leading into the Gauss formula of the next section.

Consider the function

$$F_q(s, z) = \int_0^{\infty} [\theta(t) - P_q\theta(t)]\, e^{-zt} t^s \frac{dt}{t} \tag{8}$$

and set $F = F_0$, which is especially important among the Laplace-Mellin transforms of the functions $\theta - P_q\theta$. Following the results and techniques in §1, we shall study an analytic continuation of $F_q(s, z)$ and then compare $F_q(s, z)$ with $\xi(s, z)$.

Theorem 3.5. *The function $F_q(s,z)$ has a meromorphic continuation to the region*

$$\{\mathrm{Re}(s) > -\mathrm{Re}(q)\} \times \mathbf{U}_L.$$

Proof. Note that we can write (8) as

$F_q(s,z) =$ *region of meromorphy*

$$(9) \quad \int_0^1 [\theta(t) - P_q\theta(t)]\, e^{-zt} t^s \frac{dt}{t} \qquad \mathrm{Re}(s) > -\mathrm{Re}(q),\ \text{all } z$$

$$(10) \quad + \int_1^\infty [\theta(t) - Q_N\theta(t)]\, e^{-zt} t^s \frac{dt}{t} \qquad \text{all } s,\ \mathrm{Re}(z) > -C$$

$$(11) \quad + \int_1^\infty [Q_N\theta(t) - P_q\theta(t)]\, e^{-zt} t^s \frac{dt}{t} \qquad \text{all } s,\ z \in \mathbf{U}_L.$$

For (9), the assertion of meromorphy on the right follows from condition **AS 2** and Lemma 1.3; for (10) the assertion follows by Lemma 1.4, picking N and C as in **AS 1**. As for (11), the integral is a sum of integrals of elementary functions which we now recall. First we have

$$(12) \quad \int_1^\infty e^{-\lambda_k t} e^{-zt} t^s \frac{dt}{t} = \frac{\Gamma(s)}{(z+\lambda_k)^s} - \int_0^1 e^{-\lambda_k t} e^{-zt} t^s \frac{dt}{t},$$

so that

$$\int_1^\infty Q_N\theta(t) e^{-zt} t^s \frac{dt}{t} = \sum_{k=1}^\infty \left[\frac{\Gamma(s)}{(z+\lambda_k)^s} - \int_0^1 e^{-\lambda_k t} e^{-zt} t^s \frac{dt}{t} \right].$$

For the special case, when b_p is constant, we have

$$\int_1^\infty b_p e^{-zt} t^{s+p} \frac{dt}{t} = b_p \frac{\Gamma(s+p)}{z^{s+p}} - b_p \int_0^1 e^{-zt} t^{s+p} \frac{dt}{t}$$

$$(13) \qquad = b_p \frac{\Gamma(s+p)}{z^{s+p}} - \sum_{k=0}^\infty \frac{(-z)^k}{k!} b_p \cdot \frac{1}{s+p+k},$$

and for the general case we have the expansion replacing b_p by $B_p(\partial_s)$, namely:

$$\int_1^\infty b_p(t)e^{-zt}t^{s+p}\frac{dt}{t} = B_p(\partial_s)\left[\frac{\Gamma(s+p)}{z^{s+p}}\right] - \int_0^1 b_p(t)e^{-zt}t^{s+p}\frac{dt}{t}$$

$$(14) \qquad = B_p(\partial_s)\left[\frac{\Gamma(s+p)}{z^{s+p}}\right] - \sum_{k=0}^{\infty}\frac{(-z)^k}{k!}B_p(\partial_s)\left[\frac{1}{s+p+k}\right].$$

From this, the conclusion stated above follows since the integral in (11) is simply a sum of terms of the form given in (12), (13) and (14). □

Theorem 3.6. *For any q and for all (s,z) in the region of meromorphy, we have:*

a) *In the special case,*

$$\xi(s,z) = F_q(s,z) + \sum_{\mathrm{Re}(p)<\mathrm{Re}(q)} b_p\frac{\Gamma(s+p)}{z^{s+p}};$$

b) *in the general case,*

$$\xi(s,z) = F_q(s,z) + \sum_{\mathrm{Re}(p)<\mathrm{Re}(q)} B_p(\partial_s)\left[\frac{\Gamma(s+p)}{z^{s+p}}\right].$$

Proof. For $\mathrm{Re}(s)$ large, one can interchange the sum and integral in the definition of $F_q(s,z)$ and use that

$$\int_0^\infty P_q\theta_L(t)e^{-zt}t^s\frac{dt}{t} = \sum_{\mathrm{Re}(p)<\mathrm{Re}(q)} B_p(\partial_s)\left[\frac{\Gamma(s+p)}{z^{s+p}}\right].$$

The rest follows from the definition of $\xi(s,z)$ and analytic continuation. □

By taking $q=0$ in Theorem 3.6 and writing $F_0(s,z)=F(s,z)$, we obtain the following corollary.

Corollary 3.7. *The constant term of $\xi(s,z)$ at $s=1$ is given by*

$$\mathrm{CT}_{s=1}\xi(s,z) = F(1,z) + \sum_{\mathrm{Re}(p)<0} \mathrm{CT}_{s=1}B_p(\partial_s)\left[\frac{\Gamma(s+p)}{z^{s+p}}\right]$$

where

$$F(1,z) = \int_0^\infty [\theta(t) - P_0\theta(t)]\, e^{-zt}dt.$$

The constant term on the right is obtained simply by multiplying the Laurent expansions of $\Gamma(s+p)$ and z^{-s-p} and is a universal expression, meaning an expression that depends solely only on B_p for $\mathrm{Re}(p) < 0$. A direct calculation shows that there exists a polynomial $\widetilde{B}_p$ of degree at most $\deg B_p + 1$ such that

$$\mathrm{CT}_{s=1}B_p(\partial_s)\left[\frac{\Gamma(s+p)}{z^{s+p}}\right] = z^{-p-1}\widetilde{B}_p(\log z).$$

The possible pole of $\Gamma(s+p)$ at $s=1$ accounts for the possibility of $\deg \widetilde{B}_p$ exceeding $\deg B_p$. For convenience of the reader, we present these calculations explicitly in the special case.

Let $p \in \mathbf{C}$ and write

$$\Gamma(s+p) = \sum_{k=-1}^{\infty} \frac{c_k(p+1)}{(s-1)^k}$$

and

$$z^{-p-s} = z^{-p-1}e^{(s-1)(-\log z)} = z^{-p-1}\sum_{l=0}^{\infty}\frac{(-\log z)^l}{l!}(s-1)^l.$$

Then

$$\frac{\Gamma(s+p)}{z^{s+p}} = z^{-1-p}\sum_{n=-1}^{\infty}\left[\sum_{k+l=n}\frac{c_k(p+1)(-\log z)^l}{l!}\right](s-1)^n,$$

from which we obtain the equation

$$\mathrm{CT}_{s=1}\left[\frac{\Gamma(s+p)}{z^{s+p}}\right] = [c_0(p+1) - c_{-1}(p+1)\log z]z^{-1-p}.$$

Note that if $p \notin \mathbf{Z}_{<0}$, then $c_{-1}(p+1) = 0$ and $c_0(p+1) = \Gamma(p+1)$.

Corollary 3.8. *In the special case, $R_0(1; z)$ has the integral expression*

$$\begin{aligned}&\mathrm{CT}_{s=1}\xi(s,z)\\&=F(1,z)+\sum_{\mathrm{Re}(p)<0} b_p\left[c_0(p+1)-c_{-1}(p+1)\log z\right]z^{-1-p}\end{aligned}$$

where

$$F(1,z)=\int_0^\infty\left[\theta(t)-P_0\theta(t)\right]e^{-zt}dt.$$

Example 1. Suppose we are in the spectral case and also that $\theta(t)$ is such that the principal part $P_0\theta(t)$ is simply b_{-1}/t. Then the equations in Theorem 3.6 become

$$\xi(s,z)=\frac{b_{-1}}{s-1}+\mathrm{CT}_{s=1}\xi(s,z)+O(s-1)$$

where

$$\mathrm{CT}_{s=1}\xi(s,z)=\int_0^\infty\left[\theta(t)-\frac{b_{-1}}{t}\right]e^{-zt}dt-b_{-1}\log z.$$

Using that

$$\int_0^\infty\left[e^{-zt}-e^{-t}\right]\frac{dt}{t}=-\log z,$$

we can write

$$\mathrm{CT}_{s=1}\xi(s,z)=\int_0^\infty\left[\theta(t)e^{-zt}-\frac{b_{-1}e^{-t}}{t}\right]dt.$$

In this case, the integral expression for $R_0(1;z)$ reminds one of the Gauss formula for the logarithmic derivative of the gamma function, which will be studied in the next section.

Example 2. The regularized harmonic series is simply related to the classical Selberg zeta function $Z_X(s)$ associated to a compact hyperbolic Riemann surface X. For convenience, let us briefly recall the definition of the Selberg zeta function.

Let $\theta_X(t)$ be the trace of the heat kernel corresponding to the hyperbolic Laplacian that acts on C^∞ functions on X; hence, $\theta_X(t)$ is the theta function associated to the sequence L of eigenvalues of the Laplacian. The sequence A counts multiplicities of the eigenvalues. It is well known that

$$P_0\theta(t) = \frac{b_{-1}}{t},$$

where b_{-1} is a constant that depends solely on the genus g of X, namely

$$b_{-1} = 2\pi(2g-2).$$

The logarithmic derivative of the Selberg zeta function is defined by the equation

$$Z'_X/Z_X(s) = (2s-1)\int_0^\infty [\theta_X(t) - b_{-1}k(t)]\, e^{-s(s-1)t}dt$$

where $k(t)$ is a universal function, independent of X (see [Sa 87]). From Theorem 3.6, the definition of the Selberg zeta function, and the example above, we have the relation

$$\begin{aligned}
&(2s-1)D'_L/D_L(s(s-1)) - Z'_X/Z_X(s) \\
&= (2s-1)b_{-1}\left(\int_0^\infty \left[k(t) - \frac{1}{t}\right] e^{-s(s-1)t}dt - \log(s(s-1))\right) \\
&= (2s-1)b_{-1}\left(\int_0^\infty \left[k(t)e^{-s(s-1)t} - \frac{e^{-t}}{t}\right] dt\right).
\end{aligned}$$

This relation was used in [JoL 92d] to prove that the Selberg zeta function satisifies Artin's formalism ([La 70]).

§4. Gauss Formula

Next we show that a classical formula of Gauss for Γ'/Γ can be formulated and proved more generally for the regularized harmonic series. As before, we let $P_0\theta$ denote the principal part of an asymptotic expansion at 0. Define

$$\theta_z(t) = e^{-zt}\theta(t).$$

If $\theta(t)$ satisfies the three asymptotic conditions, then it is immediate that $e^{-zt}\theta(t)$ also satisfies the three asymptotic conditions. As before, if the principal part of the theta function is

$$P_0\theta(t) = \sum_{\mathrm{Re}(p)<0} b_p(t)t^p,$$

then

$$(1) \qquad P_0\theta_z(t) = \sum_{\mathrm{Re}(p)+k<0} \frac{(-z)^k}{k!} b_p(t)t^{p+k}.$$

Note that for any complex w, we have

$$\xi_{L+z,A}(s,w) = \xi_z(s,w) = \xi(s, z+w).$$

Recall from Theorem 1.8 that

$$-\frac{\partial}{\partial z}\xi(s, z+w) = \xi(s+1, z+w),$$

so, in particular, we have

$$R(z+w) = -\partial_z \mathrm{CT}_{s=0}\xi(s, z+w) = \mathrm{CT}_{s=1}\xi(s, z+w).$$

In the spectral case, we have, by the Lerch formula,

$$R(z+w) = D_L'/D_L(z+w).$$

Finally, recall that $\mathbf{C}\langle T\rangle$ is the algebra of polynomials in T^p with arbitrary complex powers $p \in \mathbf{C}$.

With all this, we can follow the development leading to Corollary 3.8 and state the following theorem, which we call the **general Gauss formula**.

Theorem 4.1. *There is a polynomial $S_w(z)$ of degree in z*

$$\deg_z S_w < -\mathrm{Re}(p_0)$$

with coefficients in $\mathbf{C}[\log w]\langle w\rangle$ such that for any $w \in \mathbf{C}$ with $\mathrm{Re}(w) > 0$ and $\mathrm{Re}(w) > \max_k\{-\mathrm{Re}(\lambda_k + z)\}$,

$$R(z+w) = \int_0^\infty [\theta_z(t) - P_0\theta_z(t)]\, e^{-wt} dt + S_w(z)$$

In the special case, $S_w(z) \in \mathbf{C}\langle w\rangle[z] + \mathbf{C}\langle w\rangle \log w[z]$.

Proof. From Theorem 1.8, we have for sufficiently large $\mathrm{Re}(s)$ and $\mathrm{Re}(z)$, while viewing w as fixed, the equalities

$$\begin{aligned}
-\partial_z\xi(s, z+w) &= \xi(s+1, z+w) \\
&= \xi_z(s+1, w) \\
&= \int_0^\infty [\theta_z(t) - P_0\theta_z(t)]\, e^{-wt}t^{s+1}\frac{dt}{t} \\
&\quad + \int_0^\infty P_0\theta_z(t)e^{-wt}t^{s+1}\frac{dt}{t}.
\end{aligned}$$

To compute the constant term in the expansion at $s = 0$, we can substitute $s = 0$ in the Laplace-Mellin integral of $\theta_z - P_0\theta_z$, thus getting the desired integral on the right hand side of the formula in the theorem. As for the integral of $P_0\theta_z$ we can use the expression (1) for the principal part of θ_z to get

$$\int_0^\infty P_0\theta_z(t)e^{-wt}t^s dt = \sum_{\mathrm{Re}(p)+k<0} \frac{(-z)^k}{k!} B_p(\partial_s)\left[\frac{\Gamma(s+p+k+1)}{w^{s+p+k+1}}\right].$$

In fact, we obtain an explicit formula for $S_w(z)$, namely

$$S_w(z) = \sum_{\mathrm{Re}(p)+k<0} \frac{(-z)^k}{k!}\mathrm{CT}_{s=0}B_p(\partial_s)\left[\frac{\Gamma(s+p+k+1)}{w^{s+p+k+1}}\right].$$

This expression shows that $S_w(z)$ is a polynomial in z, of degree $< -\mathrm{Re}(p_0)$, with coefficients in $\mathbf{C}\langle w\rangle[\log w]$. Recall that in the special case, all B_p are constants, and, for any p, the expression

$$\mathrm{CT}_{s=0} B_p(\partial_s)\left[\frac{\Gamma(s+p+k+1)}{w^{s+p+k+1}}\right]$$

lies in $\mathbf{C}\langle w\rangle + \mathbf{C}\langle w\rangle \log w$. With this, the proof of the theorem is complete. □

Remark. A direct calculation shows that there exists a polynomial B_p^* of degree at most $\deg B_p + 1$ such that

$$\mathrm{CT}_{s=0} B_p\left(\partial_s\right)\left[\frac{\Gamma(s+p)}{z^{s+p}}\right] = z^{-p} B_p^*(\log z).$$

The possible pole of $\Gamma(s+p)$ at $s=0$ accounts for the possibility of $\deg B_p^*$ exceeding $\deg B_p$. Using the relation

$$\frac{\Gamma(s+p+k+1)}{w^{s+p+k+1}} = \frac{(s)\cdots(s+k)\Gamma(s+p)}{w^{k+1}\cdot w^{s+p}}$$

one can hope to express $S_w(z)$ in terms of the polynomials B_p^*. However, even in the special case, such an expression is quite involved.

Corollary 4.2. *For fixed $w \in \mathbf{C}$ with $\mathrm{Re}(w) > 0$ and*

$$\mathrm{Re}(w) > \max_k\{-\mathrm{Re}(\lambda_k)\},$$

and for $z \in \mathbf{C}$ with $\mathrm{Re}(z)$ sufficiently large, the integral

$$I_w(z) = \int_0^\infty \left[\theta_z(t) - P_0\theta_z(t)\right] e^{-wt} dt,$$

as a function of z, is holomorphic in z and has a meromorphic continuation to all $z \in \mathbf{C}$ with poles at the points $\lambda_k + w$ in $L + w$ and corresponding residues equal to a_k.

Proof. Immediate from Theorem 3.4(b) and Theorem 4.1. □

In the case when $a_k \in \mathbf{Z}$ for all k, this allows us to write the integral $I_w(z)$ from Corollary 4.2 as

$$I_w = H'_w/H_w$$

where H_w is a meromorphic function on $\mathbf{C}$, uniquely defined up to a constant factor. Following Theorem 2.1, we define $S_w^{\#}(z)$ to be the integral of $S_w(z)$ with zero constant term, so $S_w(z) = \partial_z S_w^{\#}(z)$. Define $H_w(z)$ by the relation

$$D(z+w) = e^{S_w^{\#}(z)} H_w(z).$$

Then $H_w(z)$ is the unique meromorphic function of z such that

$$H'_w/H_w(z) = I_w(z)$$

and

$$H_w(0) = D(w).$$

Furthermore, we have

$$D'_L/D_L(z+w) = H'_w(z)/H_w(z) + S_w(z) = I_w(z) + S_w(z).$$

In the spectral case, meaning $a_k \in \mathbf{Z}_{\geq 0}$ for all k, $H_w(z)$ is entire. The function $I_w(z)$ will be studied in §4 of [JoL 92c], from a Fourier theoretic point of view.

Example 1. We show here how the classical Gauss formula is a special case of Theorem 4.1. Let

$$L = \{n \in \mathbf{Z}_{\geq 0}\}$$

and

$$a(n) = 1 \quad \text{for all } n.$$

The theta function can be written as

$$\theta_z(t) = \sum_{n=0}^{\infty} e^{-(z+n)t} = \frac{e^{-zt}}{1-e^{-t}}.$$

The principal part is simply

$$P_0\theta_z(t) = \frac{1}{t},$$

so $p_0 = -1$ and $p_1 = 0$. This allows us to compute the polynomial $S_w(z)$ and obtain the equation

$$S_w(z) = \mathrm{CT}_{s=0}\left[\frac{\Gamma(s)}{w^s}\right] = -\gamma - \log w.$$

Therefore, by combining the Example from §2 and Theorem 4.1, we have

$$\begin{aligned} D_L'/D_L(z+w) &= -\Gamma'/\Gamma(z+w) - \gamma \\ &= \int_0^\infty \left[\frac{e^{-zt}}{1-e^{-t}} - \frac{1}{t}\right] e^{-wt}\,dt - \gamma - \log w. \end{aligned}$$

Now set $w = 1$ to get

$$-\Gamma'/\Gamma(z+1) = \int_0^\infty \left[\frac{e^{-(z+1)t}}{1-e^{-t}} - \frac{e^{-t}}{t}\right] dt,$$

which is the classical Gauss formula.

Example 2. Let (L, A) be the sequences of eigenvalues and multiplicities associated to the Laplacian that acts on smooth sections of a power of the canonical sheaf over a compact hyperbolic Riemann surface. Then one can combine Example 2 from §3, Theorem 4.1 and the Lerch formula to establish the main theorem of [DP 86] and [Sa 87], without using the Selberg trace formula. In brief, this theorem states that the Selberg zeta function $Z_X(s)$ is expressible, up to universal gamma-like functions, in terms of the regularized product $D_L(s(s-1))$.

§5. Stirling Formula

As in previous sections, we work with sequences (L, A) that satisfy the three convergence conditions **DIR 1**, **DIR 2** and **DIR 3**, and such that the associated theta series satisfies the three asymptotic conditions **AS 1**, **AS 2** and **AS 3**. With this, we consider the asymptotic behavior of certain functions when $\mathrm{Re}(z) \to \infty$, by which we mean that we allow $\mathrm{Re}(z) \to \infty$ in some sector in the right half plane. The point of such a restriction is that in such a sector, $\mathrm{Re}(z)$ and $|z|$ have the same order of magnitude, asymptotically. Specifically, we shall determine the asymptotics of

$$-\mathrm{CT}_{s=0}\xi(s,z) \quad \text{as} \quad x = \mathrm{Re}(z) \to \infty.$$

These asymptotics apply in the spectral case to

$$-\log D_L(z) = \mathrm{CT}_{s=0}\xi(s,z),$$

and, therefore, can be viewed as a generalization of the classical Stirling formula.

To begin, we will study the asymptotics of

$$\xi(s,z) = \int_0^\infty \theta(t)e^{-zt}t^s\frac{dt}{t} \quad \text{for } \mathrm{Re}(z) \to \infty.$$

Fix $\mathrm{Re}(q) > 0$ and, as before, let $P_q\theta$ be as in **AS 2**. Let us write

$$\xi(s,z) = J_1(s,z) + J_2(s,z) + J_3(s,z)$$

where

$$J_1(s,z) = \int_0^1 (\theta(t) - P_q\theta(t))e^{-zt}t^s\frac{dt}{t}, \tag{1}$$

$$J_2(s,z) = \int_1^\infty \theta(t)e^{-zt}t^s\frac{dt}{t}, \tag{2}$$

$$J_3(s,z) = \int_0^1 P_q\theta(t)e^{-zt}t^s\frac{dt}{t}. \tag{3}$$

Recall that

$$m(q) = \max\deg B_p \quad \text{for} \quad \mathrm{Re}(p) = \mathrm{Re}(q).$$

The terms in (1) and (2) are handled in the two following lemmas.

Lemma 5.1. *Let $x = \mathrm{Re}(z)$ and fix q with $\mathrm{Re}(q) > 0$. Then*

$$J_1(0,z) = O(x^{-\mathrm{Re}(q)}(\log x)^{m(q)}) \quad \textit{for } x \to \infty.$$

Proof. Recall that

$$\theta(t) - P_q\theta(t) = O(t^{\mathrm{Re}(q)}|\log t|^{m(q)}) \quad \text{for } t \to 0.$$

Since $\mathrm{Re}(q) > 0$, $J_1(s,z)$ is holomorphic at $s = 0$ and

$$J_1(0,z) = \int_0^1 (\theta(t) - P_q\theta(t))e^{-zt}\frac{dt}{t}.$$

For any complex number s_0, we have the power series expansion

$$\frac{\Gamma(s)}{x^s} = \left(\sum_{k=-1}^{\infty} c_k(s_0)(s-s_0)^k\right) \cdot \left(x^{-s_0}\sum_{l=0}^{\infty}\frac{(-\log x)^l}{l!}(s-s_0)^l\right)$$

$$= x^{-s_0}\sum_{n=-1}^{\infty}\left[\sum_{k+l=n}\frac{c_k(s_0)(-\log x)^l}{l!}\right](s-s_0)^n, \tag{4}$$

from which we have, by **AS 2**,

$$|J_1(0,z)| \ll \int_0^1 e^{-xt}t^{\mathrm{Re}(q)}|\log t|^{m(q)}\frac{dt}{t}$$

$$\leq \mathrm{CT}_{s=\mathrm{Re}(q)}\left[(\partial_s)^{m(q)}\frac{\Gamma(s)}{x^s}\right] + \int_1^{\infty} e^{-xt}t^{\mathrm{Re}(q)}(\log t)^{m(q)}\frac{dt}{t}$$

$$= x^{-\mathrm{Re}(q)}\left[\sum_{k+l=m(q)}\frac{c_k(q)(-\log x)^l}{l!}\cdot n!\right] + O(e^{-x}/x)$$

as $x \to \infty$. This completes the proof of the lemma. ☐

Lemma 5.2. *We have*

$$J_2(0,z) = O(e^{-x}/x) \quad \text{for} \quad x \to \infty.$$

Proof. From **AS 1**, $J_2(s,z)$ is holomorphic at $s = 0$ and

$$J_2(0,z) = \int_1^\infty \theta(t) e^{-zt} \frac{dt}{t},$$

from which the lemma follows, since $\theta(t) = O(e^{ct})$ for some $c > 0$ and so, for x sufficiently large,

$$|J_2(0,z)| \le K \int_1^\infty e^{-(x-c)t} dt = \frac{K}{x-c} e^{-(x-c)},$$

yielding the stated estimate. □

As for $J_3(s,z)$, recall that

$$J_3(s,z) = \sum_{\mathrm{Re}(p)<\mathrm{Re}(q)} \left[\int_0^1 b_p(t) e^{-zt} t^{s+p} \frac{dt}{t} \right].$$

Lemma 5.3. *We have*

$$J_3(0,z) = \sum_{\mathrm{Re}(p)<\mathrm{Re}(q)} \mathrm{CT}_{s=0} B_p(\partial_s) \left[\frac{\Gamma(s+p)}{z^{s+p}} \right] + O(e^{-x}/x) \quad \text{for } x \to \infty.$$

Proof. Simply write

$$\int_0^1 e^{-zt} b_p(t) t^{s+p} \frac{dt}{t} = B_p(\partial_s) \left[\frac{\Gamma(s+p)}{z^{s+p}} \right] - \int_1^\infty e^{-zt} b_p(t) t^{s+p} \frac{dt}{t},$$

and as in Lemma 5.2, note that, for any s,

$$\int_1^\infty e^{-zt} b_p(t) t^{s+p} \frac{dt}{t} = O(e^{-x}/x) \quad \text{for } x \to \infty.$$

□

Combining the above three lemmas, we have established the following theorem, which we refer to as the **generalized Stirling's formula**.

Theorem 5.4. *Suppose q is such that $\mathrm{Re}(q) > 0$. Let*

$$\mathbf{B}_q(z) = \sum_{\mathrm{Re}(p)<\mathrm{Re}(q)} \mathrm{CT}_{s=0} B_p(\partial_s) \left[\frac{\Gamma(s+p)}{z^{s+p}} \right] \in \mathbf{C}\langle z \rangle [\log z].$$

Then for $x \to \infty$,

$$\mathrm{CT}_{s=0}\xi(s,z) = \mathbf{B}_q(z) + O(x^{-\mathrm{Re}(q)}(\log x)^{m(q)}).$$

It is important to note that Theorem 5.4 shows that the asymptotics of $\mathrm{CT}_{s=0}\xi(s,z)$ as $\mathrm{Re}(z) \to \infty$ are governed by the asymptotics of $\theta(t)$ as $t \to 0$. Also, in the spectral case, the Lerch formula (Theorem 2.1) applies to give

$$-\log D_L(z) = \mathrm{CT}_{s=0}\xi_L(s,z),$$

and, hence, Theorem 5.4 determines the asymptotics of the regularized product $\log D_L(z)$ as $\mathrm{Re}(z) \to \infty$ and, consequently, the asymptotics of the the characteristic determinant

$$\log \mathbf{D}_L(z) \quad \text{as } \mathrm{Re}(z) \to \infty.$$

Remark 1. In Remark 1 of §4 we defined the polynomials B_p^* by the formula

$$\mathrm{CT}_{s=0} B_p(\partial_s) \left[\frac{\Gamma(s+p)}{z^{s+p}} \right] = z^{-p} B_p^*(\log z).$$

Using these polynomials, one can write

$$\mathbf{B}_q(z) = \sum_{\mathrm{Re}(p)<\mathrm{Re}(q)} z^{-p} B_p^*(\log z),$$

so Theorem 5.4 becomes the statement that

$$\mathrm{CT}_{s=0}\xi(s,z) = \sum_{\mathrm{Re}(p)<\mathrm{Re}(q)} z^{-p} B_p^*(\log z) + O(x^{-\mathrm{Re}(q)}(\log x)^{m(q)})$$

as $x = \mathrm{Re}(z) \to \infty$. This restatement emphasizes the point that the asymptotics of $\theta(t)$ as $t \to 0$ determine the asymptotics of $\mathrm{CT}_{s=0}\xi(s,z)$ as $\mathrm{Re}(z) \to \infty$.

To further develop the asymptotics given in Theorem 5.4, one can use the following lemma.

Lemma 5.5. *For any $p \in \mathbf{C}$, we have*

$$\mathrm{CT}_{s=0}\left[\frac{\Gamma(s+p)}{z^{s+p}}\right]$$
$$= \begin{cases} \Gamma(p)z^{-p} & p \notin \mathbf{Z}_{\leq 0} \\ [-c_{-1}(n)\log z + c_0(n)]z^{-p} & p = -n \in \mathbf{Z}_{\leq 0}, \end{cases}$$

where

$$c_{-1}(n) = \frac{(-1)^n}{n!} \quad \text{and} \quad c_0(n) = \frac{(-1)^n}{n!}(-\gamma).$$

One can prove Lemma 5.5 by using the power series expansion (4) and the formula

$$\Gamma(s+1) = s\Gamma(s) \quad \text{and} \quad \Gamma'(1) = -\gamma,$$

which comes from the expansion

$$\Gamma(s) = \frac{1}{s} - \gamma + O(s).$$

Proposition 5.6. *In the special case, we have, for* $\mathrm{Re}(q) > 0$,

$$\mathbf{B}_q(z) = \sum_{n \in \mathbf{Z}_{\geq 0}} \frac{b_{-n}(-1)^n}{n!}(-\gamma - \log z)z^n + \sum_{\substack{p \notin \mathbf{Z}_{\leq 0} \\ \mathrm{Re}(p) < \mathrm{Re}(q)}} b_p\Gamma(p)z^{-p};$$

and so

$$\mathrm{CT}_{s=0}\xi(s,z) = \sum_{n \in \mathbf{Z}_{\geq 0}} \frac{b_{-n}(-1)^n}{n!}(-\gamma - \log z)z^n +$$
$$\sum_{\substack{p \notin \mathbf{Z}_{\leq 0} \\ \mathrm{Re}(p) < \mathrm{Re}(q)}} b_p\Gamma(p)z^{-p} + O(x^{-\mathrm{Re}(q)}(\log x)^{m(q)})$$

for $\mathrm{Re}(z) \to \infty$.

Example 1. Assume L is such that

$$\theta(t) = \frac{b_{-1}}{t} + b_0 + O(t) \quad \text{for } t \to 0.$$

By taking $q = 1$, Proposition 5.6 then states that

$$\begin{aligned}\mathrm{CT}_{s=0}\xi(s,z) &= b_{-1}\, z(\log z + \gamma) - b_0(\log z + \gamma) + O(z^{-1}) \\ &= (b_{-1}z - b_0)\log z + \gamma(b_{-1}z - b_0) + O(z^{-1})\end{aligned}$$

as $\mathrm{Re}(z) \to \infty$. A particular example of this is when $L = \mathbf{Z}_{\geq 0}$. Recall that the Lerch formula (Theorem 2.1) states

$$\log \Gamma(z) = \frac{1}{2}\log 2\pi + \mathrm{CT}_{s=0}\xi(s,z) - \gamma(z - \frac{1}{2}) - z.$$

By direct computation one sees that

$$\theta(t) = \sum_{n=0}^{\infty} e^{-nt} = \frac{1}{t} + \frac{1}{2} + O(t),$$

so $b_{-1} = 1$ and $b_0 = 1/2$. Therefore,

$$\mathrm{CT}_{s=0}\xi(s,z) = (z - \frac{1}{2})\log z + \gamma(z - \frac{1}{2}) + O(|z|^{-1})$$

as $\mathrm{Re}(z) \to \infty$. From this we conclude that

$$\log \Gamma(z) = \frac{1}{2}\log 2\pi + (z - \frac{1}{2})\log z - z + O(|z|^{-1})$$

as $\mathrm{Re}(z) \to \infty$. This is the classical Stirling formula.

The reader is referred to [Vo 87] for interesting examples of the Stirling formula arising from sequences of eigenvalues associated to differential operators, and to [Sa 87] for the example of the Barnes double gamma function, which appears a factor in the functional equation of the Selberg zeta function associated to a finite volume hyperbolic Riemann surface.

As an application of the same method used to derive Stirling's formula (Theorem 5.4), we derive an asymptotic development of the integral transform

$$\phi(x) = \int_a^{\infty} e^{-xu}\mathrm{CT}_{s=0}\xi(s,u)du, \tag{5}$$

for x real, positive, and $x \to 0$. We fix $a > 0$ and

$$a > -\mathrm{Re}(\lambda_k) \quad \text{for all } k.$$

This function will appear in our work [JoL 92b], but, for now, let us simply view (5) as a special value of an incomplete Laplace-Mellin transform.

We use the decomposition of $\xi(s,u)$ as a sum

$$\xi(s,u) = J_1(s,u) + J_2(s,u) + J_3(s,u)$$

as at the beginning of this section, and deal with the integral transform of each term separately, so

$$\phi(x) = \phi_1(x) + \phi_2(x) + \phi_3(x)$$

where

$$\phi_\nu(x) = \int_a^\infty e^{-xu} \mathrm{CT}_{s=0} J_\nu(s,u) du.$$

The next lemmas lead to Theorem 5.11, which states the asymptotic behavior of $\phi(x)$ as the real variable x approaches zero from the right. We start with the transform of J_1.

Lemma 5.7. *Let $h(t)$ be a bounded measurable function on $[0,1]$ and assume that for some q with $\mathrm{Re}(q) \geq 2$,*

$$h(t) = O(t^{\mathrm{Re}(q)}) \quad \text{as } t \to 0.$$

Let

$$F(x) = \int_0^1 \frac{h(t)}{x+t} \frac{dt}{t},$$

and let $[q] = [\mathrm{Re}(q)]$. Then F is $C^{[q]-2}$ on $[0,1]$ and has the Taylor development

$$F(x) = c_0 + \cdots + c_{[q]-3} x^{[q]-3} + O(x^{[q]-2}) \text{ as } x \to 0.$$

Proof. The lemma follows directly from Taylor's theorem by differentiating under the integral sign, which is justified by the stated assumptions. □

Remark 2. Lemma 5.7 will be applied to an integral of the form

$$\int_0^1 \frac{h(t)}{x+t} e^{-(x+t)} \frac{dt}{t} = e^{-x} \int_0^1 \frac{h(t)e^{-t}}{x+t} \frac{dt}{t},$$

which is a product of e^{-x} and an integral of exactly the type considered in Lemma 5.7.

Lemma 5.8. *Let $[q] = [\mathrm{Re}(q)]$. Then the function ϕ_1 is of class $C^{[q]-2}$ and has the Taylor expansion*

$$\phi_1(x) = \int_a^\infty e^{-xu} J_1(0,u)du = g_q(x) + O(x^{[q]-3}) \text{ as } x \to 0$$

with the Taylor polynomial g_q of degree $< \mathrm{Re}(q) - 3$.

Proof. From (1) we have, by interchanging the order of integration,

$$\int_a^\infty e^{-xu} J_1(0,u)du = \int_0^1 \frac{\theta(t) - P_q\theta(t)}{x+t} e^{-a(x+t)} \frac{dt}{t}.$$

We now apply Lemma 5.7 to the function

$$h(t) = \frac{\theta(t) - P_q\theta(t)}{t} e^{-a(x+t)},$$

noting that, by **AS 2**,

$$h(t) = O(t^{\mathrm{Re}(q)-1}),$$

as $t \to 0$. With this, the proof of the lemma is complete. □

Lemma 5.9. *The function*

$$\phi_2(z) = \int_a^\infty e^{-zu} J_2(0,u)du$$

which is defined for $\mathrm{Re}(z) > 0$, *has a holomorphic extension to include a neighborhood of* $z = 0$.

Proof. We have

$$\int_a^\infty e^{-zu} J_2(0,u)du = \int_1^\infty \frac{\theta(t)}{z+t} e^{-a(z+t)} \frac{dt}{t} = e^{-az} \int_1^\infty \frac{\theta(t)e^{-at}}{z+t} \frac{dt}{t}$$

from which the stated result immediately follows. □

Having studied the transforms of J_1 and J_2, we now deal with the transform of J_3.

Lemma 5.10. *Given* p *and* B_p, *there is a function* h_p, *meromorphic in* s *and entire in* z, *to be given explicitly below, such that for* $\mathrm{Re}(z) > 0$, *we have*

$$\int_a^\infty e^{-zu} B_p(\partial_s) \left[\frac{\Gamma(s+p)}{u^{s+p}}\right] du$$
$$= B_p(\partial_s) \left[\frac{\pi z^{s+p-1}}{\sin[\pi(s+p)]}\right] + h_p(s,z),$$

and, in particular,

$$\int_a^\infty e^{-zu} \mathrm{CT}_{s=0} B_p(\partial_s) \left[\frac{\Gamma(s+p)}{u^{s+p}}\right] du$$
$$= \mathrm{CT}_{s=0} B_p(\partial_s) \left[\frac{\pi z^{s+p-1}}{\sin[\pi(s+p)]}\right] + h_p(z),$$

where

$$h_p(z) = \mathrm{CT}_{s=0} h_p(s,z).$$

Proof. It suffices to assume that $-\mathrm{Re}(s+p)$ is a large, non-integral, real number, from which the result follows by analytic continuation in s. For this, note that

$$\int_a^\infty e^{-zu} B_p(\partial_s) \left[\frac{\Gamma(s+p)}{u^{s+p}}\right] du = B_p(\partial_s) \int_a^\infty e^{-zu} \left[\frac{\Gamma(s+p)}{u^{s+p}}\right] du.$$

From this we have

$$\int_a^\infty e^{-zu}u^{1-s-p}\frac{du}{u} = \left[\frac{\Gamma(1-s-p)}{z^{1-s-p}} - \int_0^a e^{-zu}u^{1-s-p}\frac{du}{u}\right]$$

$$= \frac{\pi z^{s+p-1}}{\sin[\pi(s+p)]\Gamma(s+p)} - \sum_{k=0}^{\infty}\frac{(-z)^k}{k!}\frac{a^{1-s-p+k}}{1-s-p+k}.$$

The lemma follows by putting

$$(6) \qquad h_p(s,z) = -\sum_{k=0}^{\infty}\frac{(-z)^k}{k!}B_p(\partial_s)\left[\frac{\Gamma(s+p)a^{1-s-p+k}}{1-s-p+k}\right],$$

and

$$h_p(z) = -\sum_{k=0}^{\infty}\frac{(-z)^k}{k!}\mathrm{CT}_{s=0}\left[B_p(\partial_s)\left[\frac{\Gamma(s+p)a^{1-s-p+k}}{1-s-p+k}\right]\right],$$

□

Remark 3. A direct calculation shows that there exists a polynomial $B_p^\#$ of degree at most $\deg B_p + 1$ such that

$$(7) \qquad \mathrm{CT}_{s=0}B_p(\partial_s)\left[\frac{\pi z^{s+p-1}}{\sin[\pi(s+p)]}\right] = z^{p-1}B_p^\#(\log z).$$

The possible zero of $\sin[\pi(s+p)]$ at $s=0$ accounts for the possibility of $\deg B_p^\#$ exceeding $\deg B_p$. Further, by combining Remark 1 and Lemma 5.10, we arrive at the formula

$$\int_a^\infty e^{-zu}u^{-p}B_p^*(\log u)du = z^{p-1}B_p^\#(\log z) + h_p(z).$$

As is clear from (6), the function h_p does depend on the choice of a. We note that the power series $h_p(z)$ and the polynomials $B_p^\#$

are given by universal formulas, depending linearly on B_p^*, hence on B_p.

As is shown in the proof of Lemma 5.3, we can write

$$J_3(s,z) = \sum_{\mathrm{Re}(p)<\mathrm{Re}(q)} \left(B_p(\partial_s) \left[\frac{\Gamma(s+p)}{z^{s+p}} \right] - I_p(s,z) \right),$$

where

$$I_p(s,z) = \int_1^\infty e^{-zt} b_p(t) t^{s+p} \frac{dt}{t}.$$

By interchanging order of integration, we can write

$$\int_a^\infty e^{-uz} I_p(s,u) du = \int_1^\infty \frac{e^{-a(z+t)}}{z+t} b_p(t) t^{s+p} \frac{dt}{t}.$$

Therefore, Lemma 5.9 applies to imply the existence of a function f_p, holomorphic in a neighborhood of 0, such that

$$f_p(z) = \int_a^\infty e^{-uz} I_p(s,u) du,$$

so we can write

$$\phi_3(z) = \sum_{\mathrm{Re}(p)<\mathrm{Re}(q)} [z^{p-1} B_p^{\#}(\log z) + h_p(z) + f_p(z)].$$

With all this, we can combine Lemmas 5.8, 5.9, and 5.10 and obtain the following theorem.

Theorem 5.11. *Let $\{g_q\}$, ϕ_2, $\{B_p^{\#}\}$, $\{f_p\}$ and $\{h_p\}$ be the above sequences of polynomials and entire functions for those p for which $B_p \neq 0$. Then for each q with* $\mathrm{Re}(q) > 3$

$$\int_a^\infty e^{-tu} \mathrm{CT}_{s=0} \xi(s,u) du =$$

$$\sum_{\mathrm{Re}(p)<\mathrm{Re}(q)} \left[t^{p-1} B_p^{\#}(\log t) + h_p(t) + f_p(t) \right]$$

$$+ g_q(t) + \phi_2(t) + O(t^{[q]-3})$$

as the real variable t approaches zero from the right.

Example 2. In the case $L = \mathbf{Z}_{\geq 0}$ we have that

$$\theta(t) = \sum_{n=0}^{\infty} e^{-nt} = \frac{1}{1-e^{-t}} = \sum_{n=-1}^{\infty} b_n t^n \tag{8}$$

as t approaches zero. In particular, the sequence $\{p\}$ is simply $\mathbf{Z}_{\geq -1}$ and all polynomials B_p have degree zero. Since

$$\mathrm{CT}_{s=0}\left[\frac{\pi t^{s+n-1}}{\sin[\pi(s+n)]}\right] = (-t)^n \frac{\log t}{t},$$

we have, by using the Lerch formula (Theorem 2.1), the equation

$$\int_1^{\infty} e^{-tu} \log \Gamma(u) du = \frac{\log t}{t}\left(\sum_{n=-1}^{\infty} b_n(-t)^n\right) + h(t). \tag{9}$$

Let $q \to \infty$ in Theorem 5.11. Using the absolute convergence of (8), we get (9). In order to deduce (9) from Theorem 5.11, we have used that, upon letting q approach infinity, the power series (8) converges in a neighborhood of the origin. In general, questions concerning the convergence of the above stated power series, which are necessarily questions concerning the growth of the coefficients b_n, must be addressed. For now, let us complete this example by combining the above results to conclude that

$$\int_1^{\infty} e^{-tu} \log \Gamma(u) du = \frac{\log t}{t}\frac{1}{1-e^t} + h(t).$$

This formula verifies calculations that appear in [Cr 19] (see also [JoL 92b]).

Remark 4. The formal power series arising from the asymptotic expansion in Theorems 5.4 and 5.11 (letting $\mathrm{Re}(q) \to \infty$) are interesting beyond their truncations mod $O(t^{[q]-3})$. In important applications, and notably to positive elliptic operators, these power series are convergent, and define entire functions. This is part of the theory of Volterra operators, c.f. [Di 78], Chapter XXIII, (23.6.5.3).

§6. Hankel Formula

The gamma function and the classical zeta function are well known to satisfy a Hankel formula, that is they are representable as an integral over a Hankel contour. The existence of such a formula depends on the integrand having at least an analytic continuation over the Hankel contour, and expecially having an analytic continuation around 0. Indeed, the integrands in these classical Hankel transforms are essentially theta functions. Unfortunately, theta functions cannot always be analytically continued around 0. For instance, $\sum e^{-n^2 t}$ cannot, although $\sum e^{-nt}$ can. Thus the analogues of the classical Hankel transform representations are missing in general.

However, we shall give here one possible Hankel type formula expressing the Hurwitz xi function associated to the sequences (L, A) as a complex integral of the regularized harmonic series. Observe that the Hankel formula which we prove here is different from the classical Hankel representation of the gamma function or the zeta function. We note that such a formula was used by Deninger for the Riemann zeta function (see §3 of [De 92]).

As in previous sections, let (L, A) be sequences whose Dirichlet series satisfy the three convergence conditions **DIR 1**, **DIR 2** and **DIR 3**, and whose associated theta function satisfies the asymptotic conditions **AS 1**, **AS 2** and **AS 3**. We suppose that the sequence $L = \{\lambda_k\}$ is such that

$$\mathrm{Re}(\lambda_k) > 0 \quad \text{for all } k.$$

Since we deal with arbitrary Dirichlet series $\zeta(s) = \sum a_k \lambda_k^{-s}$, it is not the case in general that there is a regularized product whose logarithmic deriviative gives the constant term $\mathrm{CT}_{s=1}\xi(s, z)$; indeed, this exists only in the case when $a_k \in \mathbf{Z}$ for all k. However, important applications will be made to the spectral case, meaning when $a_k \in \mathbf{Z}_{\geq 0}$ for all k, in which case such a regularized product D_L exists. Thus, for this case, we record here once more the

Basic Identity:

$$R(z) = \mathrm{CT}_{s=1}\xi(s, z) = D_L'/D_L(z).$$

All the formulas of this section involving

$$R(z) = \mathrm{CT}_{s=1}\xi(s, z)$$

may then be used with $R(z)$ replaced by $D'_L/D_L(z)$ in the applications to the spectral case.

Recall from **DIR 2**, **DIR 3** and Theorem 1.8 that for any integer $n > \sigma_0$ we have the formula

$$\partial_z^n R(z) = (-1)^n \mathrm{CT}_{s=1}\xi(s+n, z) = (-1)^n \Gamma(n) \sum_{k=1}^{\infty} \frac{a_k}{(z+\lambda_k)^n}.$$

To begin, let us establish the following general lemma.

Lemma 6.1. *With assumptions as above, for every sufficiently large positive integer m, there exists a real number T_m with $m \leq T_m \leq m+1$ and a real number σ_2 such that*

$$|R(z)| = O(|z|^{\sigma_2}) \quad \text{for } |z| = T_m, \quad \text{and as} \quad T_m \to \infty.$$

Proof. For every sufficiently large positive integer m, let us write

$$\begin{aligned} \partial_z^n R(z) &= (-1)^n \Gamma(n) \sum_{|\lambda_k| \leq 2m} \frac{a_k}{(z+\lambda_k)^n} \\ &\quad + (-1)^n \Gamma(n) \sum_{|\lambda_k| > 2m} \frac{a_k}{(z+\lambda_k)^n}. \end{aligned} \tag{1}$$

If we restrict z to the annulus $m < |z| < m+1$, we have the estimate

$$\sum_{|\lambda_k|>2m} \frac{a_k}{(z+\lambda_k)^n} \leq \sum_{|\lambda_k|>2m} \frac{|a_k|}{(|\lambda_k| - m)^n} \leq 2 \sum_{|\lambda_k|>2m} \frac{|a_k|}{|\lambda_k|^n}$$

which is bounded independent of m provided $n \geq \sigma_0$.

By the convergence condition **DIR 2(b)**, we have that

$$\#\{\lambda_k : m < |\lambda_k| < m+1\} = O(m^{\sigma_1}),$$

so there are $O(m^{\sigma_1+1})$ terms in the first sum. Also, we conclude the existence a sequence $\{T_m\}$ of real numbers, tending to infinity and with $m < T_m < m+1$, such that if $|z| = T_m$, then the distance from z to any $-\lambda_k$ is at least $cm^{-\sigma_1}$ with a suitable small constant c. From **DIR 2(a)** we have that

$$|a_k| = O(|\lambda_k|^{\sigma_0}),$$

so, if $|\lambda_k| < 2m$, we have for those k for which $|\lambda| < 2m$, the bound $|a_k| = Cm^{\sigma_0}$, for a suitable large constant C. With all this, the first sum can be bounded by

$$O(m^{\sigma_0} \cdot m^{\sigma_1+1} \cdot m^{n\sigma_1}).$$

If $|z| = T_m$, we can write this bound as stating

$$\sum_{|\lambda_k|\leq 2m} \frac{a_k}{(z+\lambda_k)^n} = O(|z|^{1+\sigma_0+(n+1)\sigma_1}). \tag{2}$$

This establishes that $\partial_z^n R(z)$ has polynomial growth on the (increasing) circles centered at the origin with radius T_m. From the expression

$$\partial_z^n R(z) = (-1)^n \int_0^\infty \theta(t) e^{-zt} t^n \frac{dt}{t},$$

one has that $\partial_z^n R(z)$ is bounded for $z \in \mathbf{R}_{>0}$ with z sufficiently large. Upon integrating the function $\partial_z^n R(z)$ n times along a path consisting of $\mathbf{R}_{>0}$ and the circle $|z| = T_m$, we conclude that $R(z)$ itself has polynomial growth on the circles $|z| = T_m$. Indeed, from (2) and the fact that $\sigma_0 \geq -\mathrm{Re}(p_0)$, we have

$$|R(z)| = O(|z|^{\sigma_0+(n+1)(\sigma_1+1)}) \quad \text{for} \quad |z| = T_m, \quad \text{and as} \quad T_m \to \infty.$$

This completes the proof of the lemma. $\square$

It should noted that the proof of Lemma 6.1 explicitly constructs the value of σ_2. Indeed, since one can take $n < \sigma_0 + 1$, σ_2 can be written in terms of σ_0 and σ_1.

Having established this preliminary lemma, we can now proceed with our Hankel formula for the regularized harmonic series R. Let δ be a fixed positive number and assume

$$\delta < |\lambda_k| \quad \text{for all } \lambda_k \in L.$$

Let C_δ denote the contour consisting of:

- the lower edge of the cut from $-\infty$ to $-\delta$ in the cut plane $\mathbf{U}_L$;
- the circle S_δ, given by $w = \delta e^{i\phi}$ for ϕ ranging from $-\pi$ to π;
- and the upper edge of the cut from $-\delta$ to $-\infty$ in the cut plane $\mathbf{U}_L$.

If $G(z)$ is a meromorphic function in $\mathbf{U}_L$, then

$$\int_{C_\delta} G(w)dw = \int_{-\infty}^{-\delta} + \int_{S_\delta} + \int_{-\delta}^{-\infty} G(w)dw.$$

Symbolically, let us set

$$\int_{-\infty}^{-\delta} + \int_{-\delta}^{-\infty} = \int_{C_\delta} - \int_{S_\delta}.$$

When taking the sum of the two integrals on the negative real axis, it is of course understood that for the second integral, we deal with the analytic continuation of $G(w)$ over the circle S_δ.

We call the result of the following theorem the **Hankel formula**.

Theorem 6.2. *Let $s \in \mathbf{C}$ be such that $\mathrm{Re}(s) > \sigma_2 + 1$. Then for any z with $\mathrm{Re}(z)$ sufficiently large, we have*

$$\xi(s, z) = \frac{\Gamma(s)}{2\pi i} \int_{C_\delta} R(z - w)w^{-s}dw.$$

Proof. Let $T \in \mathbf{R}_{>0}$ be such that $|\lambda_k| \neq T$ for all k, and let $C_{\delta,T}$ denote the contour consisting of:

- the lower edge of the cut from $-T$ to $-\delta$ in $\mathbf{U}_L$;
- the circle S_δ, given by $w = \delta e^{i\phi}$ for ϕ ranging from $-\pi$ to π;
- the upper edge of the cut from $-\delta$ to $-T$ in $\mathbf{U}_L$;
- and the circle S_T, given by $w = Te^{i\phi}$ for ϕ ranging from π to $-\pi$.

Combining the residue theorem with Theorem 3.4(b), we get

$$\frac{1}{2\pi i}\int\limits_{C_{\delta,T}} R(z-w)w^{-s}dw = \sum_{0<|z+\lambda_k|<T} \frac{a_k}{(z+\lambda_k)^s}.$$

Now let $T \to \infty$ along the sequence $\{T_m\}$ that was constructed in Lemma 6.1. With this, the integral over S_T will go to zero as T approaches infinity if s is such that $\mathrm{Re}(s) > \sigma_2 + 1$. For these values of s, the limit of the sum above is the Hurwitz zeta function, and the theorem is proved. □

For the remainder of this section we will study the analytic continuation of the integral given in Theorem 6.2. The proof of Theorem 6.2 shows that the problem in extending the region for which the integral converges arises because of the asymptotic behavior of $R(z-t)$ for $-\mathrm{Re}(t)$ large. In any case, the following lemma takes care of the integral over the circle S_δ.

Lemma 6.3. *For any fixed δ sufficiently small, and any z such that $\mathrm{Re}(z)$ is sufficiently large, the integral*

$$\frac{1}{2\pi i}\int\limits_{S_\delta} R(z-w)w^{-s}dw$$

is holomorphic for all $s \in \mathbf{C}$.

Proof. Since $\mathrm{Re}(z)$ is sufficiently large, the integrand is absolutely bounded on S_δ. □

Lemma 6.3 implies that the analytic continuation of the Hankel formula to any s for which $\mathrm{Re}(s) \leq \sigma_2 + 1$ must take account the asymptotic behavior of $R(z-w)$ for fixed z and as $-\mathrm{Re}(w) \to \infty$. We can use the generalized Stirling's formula, Theorem 5.4, to extend the Hankel formula in much the same way **AS 2** is used to extend the Hurwitz xi function (see, in particular, Lemma 1.3 and Theorem 1.5).

Lemma 6.4. *For fixed $\delta > 0$, for any z with $\mathrm{Re}(z)$ sufficiently large and for any q and p for which $\mathrm{Re}(q) > \mathrm{Re}(p)$, the integral*

$$\frac{1}{2\pi i}\left(\int\limits_{-\infty}^{-\delta} + \int\limits_{-\delta}^{-\infty}\right)[R(z-t) - \mathbf{B}_q(z-t)]\,t^{-s}dt$$

is holomorphic for $\mathrm{Re}(s) = \sigma > -\mathrm{Re}(q) + 1$.

Proof. For any $T > 1$, the integral over the segments from $-T$ to $-\delta$, both the lower cut and the upper cut, is finite for all s since the integrand is absolutely bounded. By Theorem 5.4, we have

$$\left| \frac{1}{2\pi i} \left(\int_{-\infty}^{-T} + \int_{-T}^{-\infty} \right) \left[R(z-t) - \mathbf{B}_q(z-t) \right] t^{-s} dt \right|$$

$$\ll \int_{-\infty}^{-T} |t|^{-\sigma - \mathrm{Re}(q)} dt,$$

which converges if $-\sigma - \mathrm{Re}(q) < -1$, as asserted. □

Proposition 6.5. *Let* q *be such that* $\mathrm{Re}(q) > 1$, *and let* δ *be a sufficiently small positive number such that* $|\lambda_k| > \delta$ *for all* k. *Then for any* z *with* $\mathrm{Re}(z)$ *sufficiently large,*

$$\mathrm{CT}_{s=0}\xi(s,z) =$$

$$\frac{1}{2\pi i} \left(\int_{-\infty}^{-\delta} + \int_{-\delta}^{-\infty} \right) \left[R(z-t) - \mathbf{B}_q(z-t) \right] (-\gamma - \log t)\, dt$$

$$+ \mathrm{CT}_{s=0} \frac{\Gamma(s)}{2\pi i} \left(\int_{-\infty}^{-\delta} + \int_{-\delta}^{-\infty} \right) \mathbf{B}_q(z-t) t^{-s} dt$$

$$+ \frac{1}{2\pi i} \int_{S_\delta} R(w-t)(-\gamma - \log w)\, dw.$$

Proof. Write the integral over C_δ as the sum of integrals over S_δ and the line segments between $-\infty$ and $-\delta$. Since

$$\begin{aligned} \mathrm{CT}_{s=0} \frac{\Gamma(s)}{w^s} &= \mathrm{CT}_{s=0} \left(\left[\frac{1}{s} - \gamma + O(s^2) \right] \left[1 - s \log w + O(s) \right] \right) \\ &= -\gamma - \log w, \end{aligned} \tag{3}$$

we have, by Lemma 6.3,

$$\mathrm{CT}_{s=0}\frac{\Gamma(s)}{2\pi i}\int\limits_{S_\delta} R(z-w)w^{-s}dw = \frac{1}{2\pi i}\int\limits_{S_\delta} R(z-w)(-\gamma-\log w)\,dw.$$

Lemma 6.4 yields a similar result for the integral of $R - \mathbf{B}_q$. □

Next, we let δ approach zero and show that the integral over S_δ will go to zero.

Lemma 6.6. *For any q and p for which* $\mathrm{Re}(q) > \mathrm{Re}(p)$, *and for any z such that* $\mathrm{Re}(z)$ *is sufficiently large, we have*

$$\lim_{\delta\to 0}\int\limits_{S_\delta} R(z-w)\log w\; dw = 0$$

and

$$\lim_{\delta\to 0}\left[\left(\int\limits_{-\infty}^{-\delta} + \int\limits_{-\delta}^{-\infty}\right)[R(z-t)-\mathbf{B}_q(z-t)]\log t\; dt\right]$$
$$= \left(\int\limits_{-\infty}^{0} + \int\limits_{0}^{-\infty}\right)[R(z-t)-\mathbf{B}_q(z-t)]\log t\; dt.$$

Proof. If $\mathrm{Re}(z)$ is sufficiently large, the functions

$$R(z-t) \quad\text{and}\quad R(z-t)-\mathbf{B}_q(z-t)$$

are bounded as t approaches zero. The second term is bounded by $t^{-\mathrm{Re}(q)}$ as $t \to 0$. Therefore, the first integral is bounded by a multiple of

$$\left|\int\limits_{S_\delta}\log t\; dt\right| \ll \delta\log\delta,$$

which approaches zero as δ approaches zero. The same estimate proves the second assertion. □

Combining Lemma 6.4, Proposition 6.5 and Lemma 6.6, we have

Theorem 6.7. *With notation as above,*

$$\mathrm{CT}_{s=0}\xi(s,z) =$$

$$\frac{1}{2\pi i}\left(\int_{-\infty}^{0} + \int_{0}^{-\infty}\right)[R(z-t) - \mathbf{B}_q(z-t)]\,(-\gamma - \log t)\;dt$$

$$+ \lim_{\delta\to 0} \mathrm{CT}_{s=0}\frac{\Gamma(s)}{2\pi i}\left(\int_{-\infty}^{-\delta} + \int_{-\delta}^{-\infty}\right)\mathbf{B}_q(z-t)t^{-s}dt.$$

One can view the content of Theorem 6.7 as a type of regularized form of the Fundamental Theorem of Calculus. One should note the presence of the term $-\gamma - \log t$ in Theorem 6.7, which is a function that also appeared in our general Stirling formula; see Proposition 5.6 of the previous section.

To conclude, let us note that if instead of considering the Hurwitz xi function we would have studied the Hurwitz zeta function we would have obtained the following result.

Theorem 6.8. *With notation as above,*

$$\mathrm{CT}_{s=0}\zeta'(s,z) =$$

$$\frac{1}{2\pi i}\left(\int_{-\infty}^{0} + \int_{0}^{-\infty}\right)[R(z-t) - \mathbf{B}_q(z-t)]\,(-\log t)\;dt$$

$$+ \lim_{\delta\to 0} \mathrm{CT}_{s=0}\partial_s\frac{1}{2\pi i}\left(\int_{-\infty}^{-\delta} + \int_{-\delta}^{-\infty}\right)\mathbf{B}_q(z-t)t^{-s}dt.$$

The proof of Theorem 6.8 follows that of Theorem 6.7 with the only change being the use of the formula

$$\mathrm{CT}_{s=0}\partial_s w^{-s} = -\log w$$

in place of (3).

§7. Mellin Inversion Formula

So far we have considered sequences (L, A) that satisfy the three convergence conditions **DIR 1**, **DIR 2** and **DIR 3** and whose associated theta function θ satisfies the three asymptotic conditions **AS 1**, **AS 2** and **AS 3**. From these assumptions we then derived properties concerning various transforms. We now want to perform other operations, so we reconsider these axioms *ab ovo*. Especially, we shall consider the inverse Mellin transform which gives θ in terms of ζ. Throughout this section we shall assume

$$\mathrm{Re}(\lambda_k) > 0 \quad \text{for all} \quad k.$$

Recall that Theorem 1.12 proved that the convergence conditions **DIR 1**, **DIR 2** and **DIR 3** imply the asymptotic conditions **AS 1** and **AS 3**. With this, Theorem 1.11 applies to show that for $\mathrm{Re}(s)$ sufficiently large we have

$$\zeta(s) = \sum_{k=1}^{\infty} \frac{a_k}{\lambda_k^s} = \frac{1}{\Gamma(s)} \mathbf{M}\theta(s).$$

As previously stated, the asymptotic condition **AS 2** is quite delicate. In this section, we will show how, by imposing additional assumptions of meromorphy and certain growth conditions on ζ, the asymptotic condition **AS 2** follows. To begin, we need the following lemma which addresses the question of convergence of the partial series

$$\sum_{k=1}^{N-1} \frac{a_k}{\lambda_k^s},$$

to $\zeta(s)$ as we let $N \to \infty$.

Lemma 7.1. *Assume that the sequences (L, A) satisfy the convergence conditions* **DIR 1**, **DIR 2** *and* **DIR 3**, *and let σ_0 be as in* **DIR 2(a)**. *For each $N \geq 1$ let*

$$\psi_N = \sup|\arg(\lambda_k)| \quad \text{for all } k \geq N,$$

so $\psi_N < \frac{\pi}{2}$ for all N sufficiently large.

(a) *If $s = \sigma + it$ is such that*

$$\mathrm{Re}(s) = \sigma > \sigma_0,$$

then for N sufficiently large

$$\left|\zeta(s) - \sum_{k=1}^{N-1} a_k \lambda_k^{-s}\right| \leq e^{|t|\psi_N} \cdot \sum_{k=N}^{\infty} |a_k||\lambda_k|^{-\sigma}.$$

(b) *For all s in any fixed compact subset of the half plane*

$$\mathrm{Re}(s) = \sigma > \sigma_0,$$

the convergence

$$\lim_{N\to\infty} \left|\zeta(s) - \sum_{k=1}^{N-1} a_k \lambda_k^{-s}\right| = 0,$$

is uniform.

Proof. If we write $\log \lambda_k = \log|\lambda_k| + i \arg(\lambda_k)$, we have the bound

$$\begin{aligned}\left|\zeta(s) - \sum_{k=1}^{N-1} a_k \lambda_k^{-s}\right| &\leq \sum_{k=N}^{\infty} |a_k| \cdot |\lambda_k^{-s}| \\ &= \sum_{k=N}^{\infty} |a_k| \cdot |e^{-s\log\lambda_k}| \\ &\leq \sum_{k=N}^{\infty} |a_k| \cdot e^{-\sigma\log|\lambda_k| + |t|\psi_N}.\end{aligned}$$

If we define the **absolute zeta** to be

$$\zeta_{\mathbf{abs}}(\sigma) = \sum_{k=1}^{\infty} |a_k||\lambda_k|^{-\sigma}$$

and

$$\zeta_{\mathbf{abs}}^{(N)}(\sigma) = \zeta_{\mathbf{abs}}(\sigma) - \sum_{k=1}^{N-1} |a_k||\lambda_k|^{-\sigma},$$

then we have shown that

$$\left|\zeta(s) - \sum_{k=1}^{N-1} a_k \lambda_k^{-s}\right| \leq e^{|t|\psi_N} \cdot \zeta_{\text{abs}}^{(N)}(\sigma),$$

which establishes part (a). Note that if $\sigma > \sigma_0$, then, by the convergence condition **DIR 2(a)**,

$$\lim_{N\to\infty} \zeta_{\text{abs}}^{(N)}(\sigma) = 0,$$

which shows that the upper bound in (a) approaches zero as N approaches ∞, for fixed s with $\text{Re}(s)$ sufficiently large. As for (b), since ψ_N is bounded, if s lies in a compact subset K of the half plane $\text{Re}(s) > \sigma_0$, then $|t|\psi_N$ is bounded independent of N for all $s \in K$. This completes the proof of the lemma. □

Let φ be a suitable function, which will be appropriately characterized below. For any $\sigma \in \mathbf{R}$, let $\mathcal{L}(\sigma)$ be the vertical line $\text{Re}(s) = \sigma$ in $\mathbf{C}$. Under suitable conditions on φ which guarantee the absolute convergence of the following integral, we define the **vertical transform** $\mathbf{V}_\sigma\varphi$ of φ to be

$$\mathbf{V}_\sigma\varphi(t) = \frac{1}{2\pi i} \int_{\mathcal{L}(\sigma)} \varphi(s)\Gamma(s)t^{-s}\,ds.$$

From Stirling's formula (see Theorem 5.4 and, specifically, Example 1 of §5), one sees that on vertical lines $\mathcal{L}(\sigma)$, the gamma function has the decaying behavior

$$\Gamma(s) = O_c(e^{-c|s|}) \quad \text{for every } c \text{ with } 0 < c < \pi/2, \quad \text{and } |s| \to \infty, \tag{2}$$

where, as indicated, the implied constant depends on c. Previously we studied the Mellin transform and, from Theorem 1.10, we have

$$\zeta(s) = \frac{1}{\Gamma(s)}\mathbf{M}\theta(s) = \sum_{k=1}^{\infty} \frac{a_k}{\lambda_k^s},$$

where

$$\theta(t) = \sum_{k=1}^{\infty} a_k e^{-\lambda_k t}.$$

We shall now study the **inversion formula**

$$\theta(t) = \mathbf{V}_\sigma \zeta(t) = \frac{1}{2\pi i} \int_{\mathcal{L}(\sigma)} \zeta(s)\Gamma(s)t^{-s} ds, \tag{3}$$

which is valid if

$$\sigma > -\mathrm{Re}(\lambda_k) \quad \text{for all } k.$$

The inversion formula (3) is essentially a standard elementary inversion obtained for each individual term from the relation

$$e^{-u} = \frac{1}{2\pi i} \int_{\mathcal{L}(\sigma)} \Gamma(s)t^{-s} ds. \tag{4}$$

The classical proof of (4) comes from an elementary contour integration along a large rectangle going to the left, using the fact that

$$\mathrm{res}_{-n}\Gamma(s) = \frac{(-1)^n}{n!}.$$

The relation (4) is then applied by putting $u = \lambda_k t$ and summing over k. More precisely:

Proposition 7.2. *Assume that ζ satisfies* **DIR 1, DIR 2** *and* **DIR 3**, *and let σ_0 be as defined in* **DIR 2(a)**. *Then for every $\delta > 0$, the associated theta series converges absolutely and uniformly for $t \geq \delta > 0$ and*

$$\theta = \mathbf{V}_\sigma \zeta \quad \textit{for } \sigma > \sigma_0.$$

Proof. Let $g = \mathbf{V}_\sigma \zeta$ be the vertical transform of ζ with $\sigma > \sigma_0$. By (4) we have

$$g(t) - \sum_{k=1}^{N-1} a_k e^{-\lambda_k t} = \frac{1}{2\pi i} \int_{\mathcal{L}(\sigma)} \left[\zeta(s) - \sum_{k=1}^{N-1} a_k \lambda_k^{-s} \right] \Gamma(s)t^{-s} ds. \tag{5}$$

For N sufficiently large, take ψ_N as in Lemma 7.1, and let us write $s = \sigma + iu$. With this we have the bound

$$\left| g(t) - \sum_{k=1}^{N-1} a_k e^{-\lambda_k t} \right| \leq \frac{\zeta_{\text{abs}}^{(N)}(\sigma)}{2\pi} \int_{\mathcal{L}(\sigma)} e^{|u|\psi_N} |\Gamma(\sigma + iu)| t^{-\sigma} du$$

$$= \frac{\zeta_{\text{abs}}^{(N)}(\sigma)}{2\pi} t^{-\sigma} \| e^{|u|\psi_N} \Gamma(\sigma + iu) \|_{1,\sigma},$$

where

$$\| e^{|u|\psi_N} \Gamma(\sigma + iu) \|_{1,\sigma}$$

is the L^1-norm of $e^{|u|\psi_N} \Gamma(\sigma + iu)$ on the vertical line $\mathcal{L}(\sigma)$. By the convergence condition **DIR 3** and (2), we conclude there exists a constant C, independent of N, such that

$$\left| g(t) - \sum_{k=1}^{N-1} a_k e^{-\lambda_k t} \right| \leq C \zeta_{\text{abs}}^{(N)}(\sigma) t^{-\sigma}. \tag{6}$$

If $t \geq \delta > 0$, (6) can be written as

$$\left| g(t) - \sum_{k=1}^{N-1} a_k e^{-\lambda_k t} \right| \leq C \zeta_{\text{abs}}^{(N)}(\sigma) \delta^{-\sigma}.$$

If $\sigma > \sigma_0$, $\zeta_{\text{abs}}^{(N)}(\sigma)$ tends to zero when $N \to \infty$. This shows that the theta series

$$\theta_{L,A}(t) = \theta(t) = \sum_{k=1}^{\infty} a_k e^{-\lambda_k t}$$

converges absolutely and uniformly for $t \geq \delta > 0$, as asserted in the statement of the proposition. □

Remark 1. Proposition 7.2, in particular the inequality (6), shows that the three convergence conditions **DIR 1**, **DIR 2** and **DIR 3** implies the asymptotic condition **AS 3**. This provides another proof of the first part of Theorem 1.12.

Since **AS 2** was used to show that $\mathbf{M}\theta(s)$ has a meromorphic continuation, it is natural to expect some type of meromorphy condition on ζ in order to prove that the vertical transform $\mathbf{V}_\sigma\zeta$ satisfies the asymptotic condition **AS 2**. Independently of any σ, we define the **domain of V**, and denote it by $\mathrm{Dom}(\mathbf{V})$, to be the space of functions φ satisfying the following conditions:

V 1. φ is meromorphic and has only a finite number of poles in every right half plane.

V 2. $\varphi\Gamma$ is L^1-integrable on every vertical line where $\varphi\Gamma$ has no pole.

V 3. Let $\sigma_1 < \sigma_2$ be real numbers. There exists a sequence $\{T_m\}$, with $m \in \mathbf{Z}$ and

$$T_m \to \infty \text{ if } m \to \infty \quad \text{and} \quad T_m \to -\infty \text{ if } m \to -\infty,$$

such that uniformly for $\sigma \in [\sigma_1, \sigma_2]$, we have

$$(\varphi\Gamma)(\sigma + iT_m) \to 0 \text{ as } |m| \to \infty.$$

Very often one has the simpler condition:

V 3′. $\varphi(s)\Gamma(s) \to 0$ for $|s| \to \infty$ and s lying within any vertical strip of finite width.

Even though the condition **V 3′** is satisfied in many cases, it is necessary to have **V 3** as stated. The reason for condition **V 3** is that we shall let a rectangle of integration tend to infinity, and we need that the integral of $\varphi\Gamma$ on the top and bottom of the rectangle tends to 0. Throughout, it is understood that when we take the vertical transform, we select σ such that $\varphi\Gamma$ has no pole on $\mathcal{L}(\sigma)$. The vertical transform $\mathbf{V}_\sigma\varphi$ of φ depends on the choice of σ, of course, and we shall determine this dependence in a moment.

Remark 2. In the more standard cases, such as zeta functions of number fields or modular forms, or easier kinds of Selberg zeta functions, there is no difficulty in verifying the three conditions **V 1**, **V 2**, **V 3**, or, usually, **V 3′**. In fact, functions φ of this kind have usually at most polynomial growth in vertical strips, so their product with the gamma function decreases exponentially in vertical strips. In certain other interesting cases, it may be more difficult to prove this polynomial growth, and there are cases where it remains to be determined exactly what is the order of growth in vertical strips. The proof of polynomial growth depends on functional equations and Euler products in the classical cases.

We shall determine conditions under which we get **AS 2** for $\mathbf{V}_\sigma\varphi$. Immediately from the restricted growth of $\varphi\Gamma$ on vertical lines **V 1**, the asymptotic behavior of $\varphi\Gamma$ on horizontal lines **V 3**, and Cauchy's formula, we have:

Lemma 7.3. *Let $\varphi \in \mathrm{Dom}(\mathbf{V})$, and let $\sigma' < \sigma$ be such that $\varphi\Gamma$ has no pole on $\mathcal{L}(\sigma')$ and $\mathcal{L}(\sigma)$. Let $\mathcal{R}(\sigma', \sigma)$ be the strip (the infinite rectangle) defined by the inequalities*

$$\sigma' < \mathrm{Re}(s) < \sigma,$$

and let $\{-p\}$ be the sequence of poles of $\varphi\Gamma$. Then

$$\mathbf{V}_\sigma\varphi(t) = \mathbf{V}_{\sigma'}\varphi(t) + \sum_{-p\in\mathcal{R}(\sigma',\sigma)} \mathrm{res}_{-p}[\varphi(s)\Gamma(s)t^{-s}].$$

To continue, let us analyze the sum

$$\sum_{-p\in\mathcal{R}(\sigma',\sigma)} \mathrm{res}_{-p}[\varphi(s)\Gamma(s)t^{-s}]$$

which was obtained in Lemma 7.3. For this, let us write

$$t^{-s} = t^p \cdot t^{-(s+p)} = t^p \cdot \sum_{n=0}^{\infty} \frac{(-\log t)^n}{n!}(s+p)^n. \tag{7}$$

Let $d_p = -\mathrm{ord}_{-p}[\varphi\Gamma]$ and consider the Laurent expansion at $s = -p$:

$$\varphi(s)\Gamma(s) = \sum_{k=-d_p}^{\infty} c_k(s+p)^k, \tag{8}$$

so

$$\mathrm{res}_{-p}[\varphi(s)\Gamma(s)t^{-s}] = \sum_{k+n=-1} c_k \frac{(-\log t)^n}{n!} \cdot t^p.$$

Then from (7) and (8) we see immediately that there exists a polynomial B_p of degree d_p such that

$$\text{(9)} \qquad \operatorname{res}_{-p}[\varphi(s)\Gamma(s)t^{-s}] = B_p(\log t)t^p.$$

As in §1, given a sequence $\{p\}$ of complex numbers, ordered by increasing real parts that tend to infinity, and given a sequence of polynomials $\{B_p\}$ for every p in the sequence, we define

$$P_q(t) = \sum_{\operatorname{Re}(p)<\operatorname{Re}(q)} B_p(\log t)t^p.$$

As before, we put

$$m(q) = \max \deg B_p \text{ for } \operatorname{Re}(p) = \operatorname{Re}(q).$$

With this, we can combine Lemma 7.3 and (9) to obtain:

Theorem 7.4. *Let $\varphi \in \operatorname{Dom}(\mathbf{V})$, and let $\{-p\}$ be the sequence of poles of $\varphi\Gamma$. Let σ be a positive, sufficiently large, real number such that neither φ nor Γ have poles for $\operatorname{Re}(s) \geq \sigma$. Then the vertical transform $\mathbf{V}_\sigma\varphi$ of φ satisfies the asymptotic condition* **AS 2**, *or, briefly stated,*

$$\mathbf{V}_\sigma\varphi(t) \sim \sum_p B_p(\log t)t^p \quad \text{as } t \to 0.$$

Proof. If we let $s = \sigma' + it$ with σ' a large negative number, then we have the estimate

$$|\mathbf{V}_{\sigma'}\varphi(t)| \leq \frac{1}{2\pi} \int_{\mathcal{L}(\sigma')} \left|\varphi(s)\Gamma(s)t^{-s}\right| dt \leq \frac{1}{2\pi} t^{-\sigma'} \|\varphi\Gamma\|_{1,\sigma'},$$

where $\|\varphi\Gamma\|_{1,\sigma'}$ is the L^1-norm of $\varphi\Gamma$ on the line $\mathcal{L}(\sigma')$. By **V 2**, this integral is finite. Now simply combine Lemma 7.3 and (9), and let $t \to 0$ to conclude the proof. □

With all this, we can now summarize the connection between the convergence conditions that apply to sequences (L, A) and the asymptotic conditions that apply to the associated theta series.

Theorem 7.5. *Let (L, A) be sequences that satisfy the three convergence conditions* **DIR 1**, **DIR 2** *and* **DIR 3**, *and assume that the associated zeta function*

$$\zeta_{L,A}(s) = \zeta(s) = \sum_{k=1}^{\infty} a_k \lambda_k^{-s}$$

is in Dom(**V**). *Then the associated theta series*

$$\theta_{L,A}(t) = \theta(t) = \sum_{k=1}^{\infty} a_k e^{-\lambda_k t}$$

satisfies the asymptotic conditions **AS 1**, **AS 2** *and* **AS 3**. *The asymptotic expansion of* **AS 2** *is given by (9) and Theorem 7.4 with $\varphi = \zeta$.*

Proof. Theorem 1.12 shows that if (L, A) satisifies the three convergence conditions, then the associated theta series satisfies the asymptotic conditions **AS 1** and **AS 3**. Theorem 7.4 shows that the meromorphy assumption on ζ, namely that ζ is in Dom(**V**), implies **AS 2**. □

To conclude this section, we will use Theorem 7.4 to show how, given sequences whose theta series satisfies the asymptotic conditions, one can construct a new sequence with the same property. The following theorem follows immediately from the definitions and the holomorphy of the exponential function.

Theorem 7.6. *Let $z \in \mathbf{C}$ be fixed and assume that (L, A) is such that the associated theta series satisifies the three asymptotic conditions* **AS 1**, **AS 2** *and* **AS 3**. *Then the sequence $(L + z, A)$ satisfies the three asymptotic conditions* **AS 1**, **AS 2** *and* **AS 3**.

Proof. The theta function associated to $(L + z, A)$ is simply $e^{-zt}\theta_{L,A}(t)$, hence the proof follows.

Theorem 7.7. *Let $r > 0$ and assume that (L, A) is such that the associated theta series satisifies the three asymptotic conditions* **AS 1**, **AS 2** *and* **AS 3**. *Then the theta series associated to*

$$(\{\lambda_k^{1/r}\}, A) = (L^{1/r}, A)$$

satisfies satisfies the three asymptotic conditions **AS 1**, **AS 2** *and* **AS 3**. *If the exponents $\{p\}$ in the asymptotic expansion of $\theta_{L,A}$ near $t = 0$ are real, then so are the exponents in the asymptotic expansion of $\theta_{L^{1/r},A}$ near $t = 0$.*

Proof. Since $\zeta(s)$ is meromorphic as a function of s, so is $\zeta(rs)$, and we can apply Theorem 7.4. The reality statement is immediate from the proof of Lemma 7.3.

Let us consider two zeta functions ζ_1 and ζ_2 corresponding to the sequences (L_1, A_1) and (L_2, A_2), respectively. Let (L_3, A_3) be the **tensor product**, which we define so that L_3 is the family of all products

$$L_3 = \{\lambda_k \lambda'_j\} \quad \text{with} \quad \lambda_k \in L_1 \quad \text{and} \quad \lambda'_j \in L_2;$$

while A_3 is the family of all products $\{a_k a'_j\}$, so then the zeta function ζ_3 is simply the product

$$\zeta_3 = \zeta_1 \zeta_2.$$

When written in full, the zeta function ζ_3 is

$$\zeta_3(s) = \sum_{k,j}^{\infty} \frac{a_k a'_j}{(\lambda_k \lambda'_j)^s},$$

and the associated theta series θ_3 reads

$$\theta_3(t) = [\theta_1 \otimes \theta_2](t) = \sum_{k,j} (a_k a'_j) e^{-(\lambda_k \lambda'_j)t}.$$

To study ζ_3 and θ_3, we shall apply Lemma 7.1. For this, we define the **truncated zeta function** to be

$$\zeta^{(N)}(s) = \sum_{k=N}^{\infty} a_k \lambda_k^{-s} = \zeta(s) - \sum_{k=1}^{N-1} a_k \lambda_k^{-s}.$$

Theorem 7.8. *Let (L_1, A_1) and (L_2, A_2) be sequences such that the associated theta series satisify the three asymptotic conditions* **AS 1**, **AS 2** *and* **AS 3**. *Assume there exists an $\epsilon > 0$ such that for all j and k, we have*

$$-\frac{\pi}{2} + \epsilon \leq \arg(\lambda_k) + \arg(\lambda'_j) \leq \frac{\pi}{2} - \epsilon,$$

or, in other words, the tensor product series satisfies **DIR 3**. *Further, assume there exists some N such that the truncated zeta functions*

$$\zeta_1^{(N)}, \zeta_2^{(N)}, \quad \text{and the product} \quad \zeta_1^{(N)}\zeta_2^{(N)}$$

are in Dom(**V**). *Then the theta series associated to the tensor product (L_3, A_3) satisfies satisfies the three asymptotic conditions* **AS 1**, **AS 2** *and* **AS 3**. *If the exponents in the asymptotic expansion* **AS 2** *are real for θ_1 and θ_2, then they are also real for $\theta_3 = \theta_1 \otimes \theta_2$.*

Proof. Since θ_1 and θ_2 satisfy **AS 1**, it is immediate that θ_3 satisfies **AS 1**. Also, it is immediate that the tensor product (L_3, A_3) satisifes the convergence conditions **DIR 1** and **DIR 2** since the product of two absolutely convergent Dirichlet series is absolutely convergent. Therefore, by Corollary 1.10, we know that the zeta functions ζ_1 and ζ_2 are meromorphic in **C**. By assuming that the tensor product series satisifies **DIR 3**, Proposition 7.2 applies to show that $\theta_3 = \theta_1 \otimes \theta_2$ satisifes **AS 3**. Finally, by assuming that the truncated zeta functions are in Dom(**V**), we may apply Theorem 7.4 to conclude that θ_3 satisfies **AS 2**.

In the case of real exponents for the asymptotic expansions of θ_1 and θ_2, we know, again by Corollary 1.10, that the poles of ζ_1 and ζ_2 are of the form $-(p+n)$ with real p and $n \in \mathbf{Z}$. Hence, by the proof of Lemma 7.3, the exponents of the asymptotic expansion for θ_3, which are the poles of ζ_3, are also real, thus concluding proof of the theorem. □

Example 1. Let L be the sequence of eigenvalues associated to a Laplacian that acts on C^∞ functions on a compact hyperbolic Riemann surface X. The parametrix construction of the heat kernel shows that L satisfies the three asymptotic conditions **AS 1**, **AS 2** and **AS 3**. From Theorem 7.6 we have that $L - 1/4$ satisfies the three asymptotic conditions, and Theorem 7.7 implies that the sequence

$$\sqrt{L - 1/4} = \{\sqrt{\lambda_k - 1/4}\} = \{r_k\}$$

also satisfies the three asymptotic conditions. The sequence

$$1/2 + \sqrt{-1} \cdot \sqrt{L - 1/4} = \{1/2 + ir_k\}$$

is precisely the set of zeros $\{\rho\}$ with $\mathrm{Im}(\rho) > 0$ for the Selberg zeta function associated to X. Note that the general Cramér theorem proved in [JoL 92b] also proves that the sequence $\sqrt{L - 1/4}$ satisfies the three asymptotic conditions **AS 1**, **AS 2** and **AS 3**.

Essentially the same argument, applied in reverse, holds for general zeta functions, such as those associated to the theta function as in Cramér's theorem ([JoL 92b]).

Example 2. The Dedekind zeta function of a number field

$$\zeta(s) = \sum_{\mathfrak{a}} \mathrm{N}\mathfrak{a}^{-s} = \sum_{k=1}^{\infty} a_k k^{-s}$$

satisfies all three conditions **DIR 1**, **DIR 2**, and **DIR 3**, and lies in Dom(**V**). Here a_k is the number of ideals $\mathfrak{a}$ with $\mathrm{N}\mathfrak{a} = k$. Therefore, by Theorem 7.5, the associated theta function

$$\theta(t) = \sum_{k=1}^{\infty} a_k e^{-kt}$$

satisfies the three asymptotic conditions, especially **AS 2**. Note that this theta function is different from the theta function used in the classical (Hecke) proof of the functional equation. A similar remark of course holds for the L-series.

TABLE OF NOTATION

Because of an accumulation of conditions and notation, and their use in the current series of papers, we tabulate here the main objects and conditions that we shall consider.

We let $L = \{\lambda_k\}$ and $A = \{a_k\}$ be sequences of complex numbers. To these sequences we associate various objects.

A **Dirichlet series** or **zeta function:**

$$\zeta_{A,L}(s) = \zeta(s) = \sum_{k=1}^{\infty} a_k \lambda_k^{-s},$$

and, more generally, for each positive integer N, the **truncated Dirichlet series**

$$\zeta_{A,L}^{(N)}(s) = \zeta^{(N)}(s) = \sum_{k=N}^{\infty} a_k \lambda_k^{-s}.$$

The sequences L and A (or Dirichlet series) may be subject to the following conditions:

DIR 1. For every positive real number c, there is only a finite number of k such that $\mathrm{Re}(\lambda_k) \leq c$.

DIR 2.

(a) The Dirichlet series

$$\sum_k \frac{a_k}{\lambda_k^{\sigma}}$$

converges absolutely for some real σ. Equivalently, we can say that there exists some $\sigma_0 \in \mathbf{R}_{>0}$ such that

$$|a_k| = O(|\lambda_k|^{\sigma_0}) \text{ for } k \to \infty.$$

(b) The Dirichlet series

$$\sum_k \frac{1}{\lambda_k^{\sigma}}$$

converges absolutely for some real σ. Specifically, let σ_1 be a real number for which

$$\sum_k \frac{1}{|\lambda_k|^{\sigma_1}} < \infty.$$

We define m_0 to be the smallest *integer* ≥ 1 such that

$$\sum_{1}^{\infty} \frac{|a_k|}{|\lambda_k|^{m_0}} < \infty.$$

DIR 3. There is a fixed $\epsilon > 0$ such that for all k sufficiently large, we have

$$-\frac{\pi}{2} + \epsilon \leq \arg(\lambda_k) \leq \frac{\pi}{2} - \epsilon.$$

Equivalently, there exists positive constants C_1 and C_2 such that for all k,

$$C_1|\lambda_k| \leq \operatorname{Re}(\lambda_k) \leq C_2|\lambda_k|.$$

A **theta series** or **theta function**:

$$\theta_{A,L}(t) = \theta(t) = a_0 + \sum_{k=1}^{\infty} a_k e^{-\lambda_k t},$$

a **reduced theta series**

$$\theta_{L,A}^{(1)}(t) = \theta^{(1)}(t) = \sum_{k=1}^{\infty} a_k e^{-\lambda_k t},$$

and, more generally, for each positive integer N, the **truncated theta series**

$$\theta_{L,A}^{(N)}(t) = \theta^{(N)}(t) = \sum_{k=N}^{\infty} a_k e^{-\lambda_k t}.$$

The **asymptotic exponential polynomials** for integers $N \geq 1$:

$$Q_N(t) = a_0 + \sum_{k=1}^{N-1} a_k e^{-\lambda_k t}.$$

We are also given a sequence of complex numbers

$$\{p\} = \{p_0, \dots, p_j, \dots\}$$

with

$$\mathrm{Re}(p_0) \leq \mathrm{Re}(p_1) \leq \dots \leq \mathrm{Re}(p_j) \leq \dots$$

increasing to infinity. To every p in this sequence, we associate a polynomial B_p and we set

$$b_p(t) = B_p(\log t).$$

We then define:

The **asymptotic polynomials at** 0:

$$P_q(t) = \sum_{\mathrm{Re}(p)<\mathrm{Re}(q)} b_p(t)t^p.$$

We define

$$m(q) = \max \deg B_p \quad \text{for} \quad \mathrm{Re}(p) = \mathrm{Re}(q)$$

and

$$n(q) = \max \deg B_p \quad \text{for} \quad \mathrm{Re}(p) < \mathrm{Re}(q).$$

Let $\mathbf{C}\langle T\rangle$ be the algebra of polynomials in T^p with arbitrary complex powers $p \in \mathbf{C}$. Then, with this notation, $P_q(t) \in \mathbf{C}[\log t]\langle t\rangle$.

A function f on $(0, \infty) = \mathbf{R}_{>0}$ may be subject to **asymptotic conditions:**

AS 1. Given a positive number C and $t_0 > 0$, there exists N and $K > 0$ such that

$$|f(t) - Q_N(t)| \leq K\epsilon^{-Ct} \text{ for } t \geq t_0.$$

AS 2. For every q, we have

$$f(t) - P_q(t) = O(t^{\mathrm{Re}(q)}|\log t|^{m(q)}) \text{ for } t \to 0.$$

We will write **AS 2** as

$$f(t) \sim \sum_p b_p(t) t^p.$$

AS 3. Given $\delta > 0$, there exists an $\alpha > 0$ and a constant $C > 0$ such that for all N and $0 < t \leq \delta$ we have

$$|\theta(t) - Q_N(t)| \leq C/t^\alpha.$$

Given a polynomial B_p a direct calculation shows that there exist polynomials B_p^*, $\widetilde{B}_p$ and $B_p^\#$ of degree $\leq \deg B_p + 1$ such that:

$$\mathrm{CT}_{s=0} B_p(\partial_s) \left[\frac{\Gamma(s+p)}{z^{s+p}} \right] = z^{-p} B_p^*(\log z)$$

$$\mathrm{CT}_{s=1} B_p(\partial_s) \left[\frac{\Gamma(s+p)}{z^{s+p}} \right] = z^{-p-1} \widetilde{B}_p(\log z)$$

$$\mathrm{CT}_{s=0} B_p(\partial_s) \left[\frac{\pi z^{s+p-1}}{\sin[\pi(s+p)]} \right] = z^{p-1} B_p^\#(\log z)$$

The possible pole of $\Gamma(s+p)$ at $s = 0$ or $s = 1$ and the possible zero of $\sin[\pi(s+p)]$ at $s = 0$ accounts for the possibility of $\deg B_p^*$, $\deg \widetilde{B}_p$, or $\deg B_p^\#$ exceeding $\deg B_p$.

Part II

A Parseval Formula for Functions with a Singular Asymptotic Expansion at the Origin

Introduction

We shall determine the Fourier transform of a fairly general type of function φ which away from the origin has derivatives that are of bounded variation on $\mathbf{R}$ and are in $L^1(\mathbf{R})$, but at the origin the function has a principal part which is a generalized polynomial in the variable x and $\log x$, namely

$$\varphi(x) = \sum_p B_p(\log x)x^p + O\left(|\log x|^m\right),$$

where $\{p\}$ ranges over a finite number of complex numbers with $\mathrm{Re}(p) < 0$, for all p, B_p is a polynomial, and m is some positive integer. Thus the Fourier transform is determined as a distribution, and more generally as a functional on a large space of test functions also to be described explicitly. Alternatively, we may say that we are proving the Parseval formula for such a function φ and its Fourier transform.

Aside from the Parseval formula having intrinsic interest in pure Fourier analysis, it arises in a natural way in analytic number theory, in the theory of differential and pseudo differential operators, and more generally in the theory of regularized products as developed in [JoL 92a].

In the so-called "explicit formulas" of number theory, one proves essentially that the sum of a suitable function taken over the primes is equal to the sum of the Fourier-Mellin transform taken over the zeros of the zeta function. Classically, only very special cases were given (see Ingham [In 32]), and Weil was the first to observe that the formula was valid on a rather large space of test functions, and could be expressed as an equality of functionals. The sum over the primes includes a term at infinity, which amounts to an integral of the test function against the logarithmic derivative of the gamma function, taken over a vertical line $\mathrm{Re}(s) = a$ [We 52]. Weil proved what amounted to a Parseval formula, by determining what amounted to the Fourier transform of the logarithmic derivative of the gamma function on a vertical line in an explicit form. Weil's functional was reproduced with some additional details (making use of some general results concerning general Schwarz distributions) in [La 70]. However, the form in which Weil (and [La 70]) left the functional at infinity still required what appeared as too complicated arguments to identify it with the classical forms in the classical special cases. Barner reformulated the Weil functional in a more practical form [Ba 81], [Ba 90] and also extended the domain of validity of the formula. Our Parseval formula includes as a special case the formulas of Weil and Barner for the gamma function.

In spectral theory or in the theory of regularized series and products, there arises a regularized harmonic series R and a regularized product (or regularized determinant) D. As a corollary of the Parseval formula, we determine the Fourier transform of R, and of the logarithmic derivative $D'/D(a + it)$ on vertical lines of the form $a + it$, where D is the regularized product from §3 of [JoL 92a]. This determination is applied in our general version of explicit formulas [JoL 93]. Several other applications will also be given in subsequent papers, including for instance functional equations for general theta functions.

As to the proof, we shall first give a special case which covers the classical case of number theory and Barner's formulation. In this special case, the Parseval formula involves only the Dirac functional applied to the test function, whereas the general case concerning arbitrary regularized determinants involves higher derivatives of the test function. These higher derivatives come from the polar part in the asymptotic expansion of the theta function at 0. The special case occurs when this polar part consists only of $1/x$.

Our proof in the special case is more direct than Barner's or Weil's, and our formulation of the result already exhibits some general principles which were not immediately apparent in previous proofs. In particular, we avoid what now appear as detours by formulating lemmas in pure Fourier analysis showing how the singularity behaves under Fourier transform. These lemmas are used in the special and general case. We are indebted to Peter Jones for a proof of one of these lemmas, which has independent interest in general Fourier analysis.

§1. A Theorem on Fourier Integrals

This section is preliminary, and proves some lemmas on Fourier integrals which will be used in the next section.

We recall the **Fourier transform**

$$f^\wedge(t) = \frac{1}{\sqrt{2\pi}} \int_{-\infty}^{\infty} f(x)e^{-itx}\,dx.$$

We shall be concerned with Fourier inversion. For $f \in L^1(\mathbf{R})$ and $A > 0$ we define

$$f_A(x) = \frac{1}{\pi} \int_{-\infty}^{\infty} f(y)\frac{\sin A(x-y)}{x-y}\,dy = \frac{1}{\sqrt{2\pi}} \int_{-A}^{A} f^\wedge(t)e^{itx}\,dt.$$

The middle expression with the sine comes from the last expression and the definition of $f^\wedge$, after an application of Fubini's theorem and the evaluation of a simple integral of elementary calculus. Let

$$f^-(x) = f(-x).$$

We are interested in seeing how f_A converges to f, that is we want the inversion formula $f^{\wedge\wedge} = f^-$ in the form

$$f(x) = \lim_{A\to\infty} f_A(x).$$

We shall give conditions under which the inversion formula is true.

We let the **Schwartz space** $\mathrm{Sch}(\mathbf{R})$ be the vector space of functions which are infinitely differentiable, and such that the function and all its derivatives tend rapidly to 0 at infinity. That f **tends rapidly to 0 at infinity** means that for all polynomials P the function Pf is bounded. Then the Schwartz space is self dual, that is

$$\mathrm{Sch}(\mathbf{R})^\wedge = \mathrm{Sch}(\mathbf{R}).$$

An elementary result of analysis asserts that the formula $f^{\wedge\wedge} = f^-$ is true for f in the Schwartz space. We shall assume such an elementary result, and extend it to functions which are less smooth, namely we shall deal with functions of bounded variation. All the background material needed is contained in [La 93]. We let $\mathrm{BV}(\mathbf{R})$

denote the space of complex valued functions of bounded variation on $\mathbf{R}$, i.e. the space of functions of bounded variation on each finite interval $[a,b]$, and such that the total variations are uniformly bounded for all $[a,b]$, in other words there exists $B > 0$ such that

$$V(f,a,b) \leq B \text{ for all } [a,b].$$

We let

$$V_{\mathbf{R}}(f) = \sup_{[a,b]} V(f,a,b).$$

Remark. *If $f \in \mathrm{BV}(\mathbf{R}) \cap L^1(\mathbf{R})$ then $f(x) \to 0$ as $x \to \pm\infty$.*

Indeed, if $|f(x)| \geq c > 0$ for infinitely many x tending to infinity, since $f \in L^1(\mathbf{R})$ there are infinitely many y such that $|f(y)| \leq c/2$ (say), and so the function could not be of bounded variation.

We let $d\mu_f$ be the Riemann-Stieltjes measure associated with f, and sometimes abbreviate $d\mu_f$ by df.

The function f_A exhibits different behavior near 0 and at infinity. Its behavior will be described in part in the lemmas below. We shall need the function

$$S(x) = \int_0^x \frac{\sin t}{t} dt,$$

so S is continuous and bounded on $\mathbf{R}$. In fact, S is bounded by the area under the first arch of $(\sin t)/t$ (between 0 and π), as follows at once by the alternating nature of the integrand.

Lemma 1.1. *Let $f \in \mathrm{BV}(\mathbf{R}) \cap L^1(\mathbf{R})$. Then for $A > 0$ the function f_A is bounded, and in fact there is a uniform bound, independent of A:*

$$\|f_A\|_\infty \leq \frac{1}{\pi} \|S\|_\infty V_{\mathbf{R}}(f),$$

where $\| \quad \|$ is the sup norm.

Proof. Integrating by parts yields

$$f_A(x) = \int_{-\infty}^{\infty} f(x-t) \frac{d}{dt} S(At)dt$$

$$= - \int_{-\infty}^{\infty} S(At)df(t-x).$$

The desired bound follows by the standard absolute estimates. We have used here that over a finite interval $[a, b]$, the two terms $f(b)S(Ab)$ and $f(a)S(Aa)$ tend to 0 as $a \to -\infty, b \to \infty$ because S is bounded, and f tends to 0. This concludes the proof. □

We recall a classical theorem from Fourier analysis giving natural conditions under which Fourier inversion holds, especially at a discontinuity. For this purpose, we shall say that a function f is **normalized at a point** x if

$$f(x) = \frac{1}{2}[f(x+) + f(x-)].$$

Thus the right and left limits of f exist at x, and the value of f at x is the midpoint. We say that f is **normalized** if f is normalized at every $x \in \mathbf{R}$.

Theorem 1.2. *Let $f \in \mathrm{BV}(\mathbf{R}) \cap L^1(\mathbf{R})$, and suppose f is normalized. Then f_A is bounded independently of A and*

$$\lim_{A\to\infty} f_A(x) = f(x) \text{ for all } x \in \mathbf{R}.$$

For a proof, see Titchmarsh [Ti 48].

The remainder of this section is devoted to examining the uniformity of the convergence in Theorem 1.2. To begin, we mention a very special case.

Lemma 1.3. *There exists a function $\alpha \in \mathrm{Sch}(\mathbf{R})$ such that $\alpha^\wedge$ has compact support, and $\alpha(0) \neq 0$. For such a function, we have*

$$\alpha_A = \alpha$$

for all A sufficiently large.

Proof. Let $\beta \in C_c^\infty(\mathbf{R})$ be an even function ≥ 0 with $\beta(0) > 0$. Let $\alpha = \beta^\wedge$. Then $\beta = \alpha^\wedge$ has compact support, and the direct evaluation of the Fourier integral together with the definition of α_A shows that the other conditions are satisfied. □

The following quantitative formulation of the Riemann-Lebesgue lemma is proved by Barner in [Ba 90], Satz 82 in §21, to which the reader is referred for a proof.

Lemma 1.4. *Assume:*

(a) $g \in \mathrm{BV}(\mathbf{R})$.

(b) $g(x) = O(|x|^{\epsilon})$ *for some* $\epsilon > 0$ as $x \to 0$.

Then the improper integral that follows exists for $A > 0$ *and satisfies the bound*

$$\int_0^{\infty} g(y) e^{iAy} \frac{dy}{y} = O_g(A^{-\frac{\epsilon}{1+\epsilon}}).$$

Also, we need the following elementary lemma.

Lemma 1.5. *For all* $0 < a < b$ *and* $A > 0$ *we have*

$$\left| \int_a^b \frac{\sin At}{t} \, dt \right| \leq \frac{3}{Aa}.$$

Proof. This follows from the change of variables $u = At$, the alternating nature of the integrand, and an elementary area estimate. □

The main result of this section is the following uniform version of Theorem 1.2.

Theorem 1.6. *Let* $g \in \mathrm{BV}(\mathbf{R}) \cap L^1(\mathbf{R})$ *and assume*

$$g(x) = O(|x|^{\epsilon}) \quad \text{for } |x| \to 0.$$

Let $\delta = \min(\frac{1}{8}, \frac{\epsilon}{16})$. *Then for all* $A > 1$,

$$g_A(x) - g_A(0) = O_g(|x|^{\delta}) \quad \text{for } |x| \to 0,$$

the estimate on the right being independent of A.

Theorem 1.6 will be proved using a series of lemmas. Before continuing, let us state the following corollary of Theorem 1.6 that further refines the uniformity of the pointwise convergence result stated in Theorem 1.2.

Corollary 1.7. *Assume g has M derivatives and all the functions $g, g^{(1)}, \ldots, g^{(M)}$ are in* $\mathrm{BV}(\mathbf{R}) \cap L^1(\mathbf{R})$. *Assume that*

$$g(x) = O(|x|^{M+\epsilon}) \quad \text{for } |x| \to 0.$$

Then:

(a) *The function g_A has M derivatives $(g_A)^{(1)}, \ldots, (g_A)^{(M)}$ and*

$$(g_A)^{(k)} = \left(g^{(k)}\right)_A \qquad \text{for } k = 1, \ldots M,$$

so, without ambiguity, we can write the derivatives of g_A as $g_A^{(1)}, \ldots, g_A^{(M)}$.

(b) *We have*

$$g_A(x) - \sum_{k=0}^{M} g_A^{(k)}(0) \frac{x^k}{k!} = O(|x|^{M+\delta}) \quad \text{for } |x| \to 0,$$

the estimate on the right being independent of A.

Proof. Let h be a differentiable function on $\mathbf{R}$, with derivative h', such that $h, h' \in \mathrm{BV}(\mathbf{R}) \cap L^1(\mathbf{R})$. Since $d/du((\sin Au)/u)$ is bounded, one can interchange derivative and integral (see page 357 of [La 83]). These calculations yield that

$$\begin{aligned}
\frac{d}{dx} h_A(x) = h_A^{(1)}(x) &= \frac{d}{dx} \int_{-\infty}^{\infty} h(y) \frac{\sin A(y-x)}{(y-x)} dy \\
&= \int_{-\infty}^{\infty} h(y) \frac{d}{dx} \left[\frac{\sin A(y-x)}{(y-x)} \right] dy \\
&= \int_{-\infty}^{\infty} h(y) \left(-\frac{d}{dy} \right) \left[\frac{\sin A(y-x)}{(y-x)} \right] dy \\
&= \int_{-\infty}^{\infty} \frac{d}{dy} h(y) \left[\frac{\sin A(y-x)}{(y-x)} \right] dy \\
&= \left[\frac{d}{dx} h \right]_A (x).
\end{aligned}$$

The integration by parts step is valid since $(\sin Au)/u$ is bounded and $h(x)$ approaches zero as x approaches infinity. This proves (a) by induction, letting $h = g^{(k)}$. As to (b), if we apply Theorem 1.6 to the function $g_A^{(M)}(x)$ we get

$$g_A^{(M)}(x) - g_A^{(M)}(0) = O_g(|x|^\delta).$$

By repeatedly integrating this equation from 0 to x and applying (a), we get (b), thus proving the corollary. □

The remainder of this section is devoted to the proof of Theorem 1.6. To do so, we will write

$$g_A(x) = \int_{-1}^{1} g(y)\frac{\sin A(x-y)}{x-y}dy + \left(\int_{-\infty}^{-1} + \int_{1}^{\infty}\right) g(y)\frac{\sin A(x-y)}{x-y}dy,$$

and investigate the finite and infinite intervals separately. For notational simplicity, we will consider the integrals over the intervals $[0,1]$ and $[1,\infty)$, with the analysis over the intervals $(-\infty,-1]$ and $[-1,0]$ being identical to that over the corresponding positive intervals. We begin with the analysis over the finite intervals.

Proposition 1.8. *Let $g \in \mathrm{BV}(\mathbf{R}) \cap L^1(\mathbf{R})$. Assume in addition that there exists $\varepsilon > 0$ such that*

$$g(x) = O(|x|^\varepsilon) \text{ for } x \to 0.$$

Then there is $\delta > 0$ such that for all $A \geq 1$ we have

$$\int_0^1 g(y)\left[\frac{\sin Ay}{y} - \frac{\sin A(x-y)}{x-y}\right]dy = O_g(|x|^\delta) \text{ for } x \to 0,$$

the estimate on the right being independent of A.

Proof. We owe the proof of this proposition to Peter Jones, who showed that one can take $\delta = \min(\frac{1}{8}, \frac{\varepsilon}{16})$, as stated in Theorem 1.6. We need to split the integral over various intervals, depending on A and x. We first settle the easiest case.

Case I. *Suppose* $|x| < A^{-4}$ *and, say,* $x > 0$. *Then*

$$\left| \int_0^1 g(y) \left[\frac{\sin Ay}{y} - \frac{\sin A(x-y)}{x-y} \right] dy \right| \ll \|g\|_1 x^{1/2}.$$

Proof. Since

$$\left| \frac{d}{dy} \left(\frac{\sin Ay}{y} \right) \right| \ll A^2,$$

we use the Mean Value Theorem and the hypothesis $x < A^{-4}$ to get the bound

$$\left| \frac{\sin Ay}{y} - \frac{\sin A(x-y)}{x-y} \right| \ll A^2 x \leq x^{1/2},$$

and then we estimate the desired integral in the coarsest way with the sup norm to conclude the proof of the present case. □

Case II. *Suppose* $|x| \geq A^{-4}$, *so* $A^{-1} \leq |x|^{1/4}$.

In this case, we will bound each separate integral without using the difference of sines. Lemma 1.4 takes care of the term with $(\sin Ay)/y$, giving a bound of $x^{\epsilon/8}$. The next lemma takes care of the term with $(\sin A(x-y))/(x-y)$.

Lemma 1.9. *With the implied constant in* $\ll$ *depending on* $\|g\|_1$, $\|g\|_\infty$, *and* $V_{\mathbf{R}}(g)$, *we have for* $x \geq A^{-4}$:

$$\left| \int_0^1 g(y) \frac{\sin A(x-y)}{x-y} dy \right| \ll x^{1/8} + x^{\varepsilon/16},$$

the estimate on the right independent of A.

Proof. We split the integral

$$\int_0^1 g(y) \frac{\sin A(x-y)}{x-y} dy$$

$$= \int_{|x-y| \leq A^{-1}} + \int_{A^{-1} \leq |x-y| \leq A^{-1}x^{-1/8}} + \int_{A^{-1}x^{-1/8} \leq |x-y|}$$

$$= I_1 + I_2 + I_3.$$

For the first integral I_1, we change variable putting $u = x - y$. Then the limits of integration for u are $|u| \leq A^{-1}$, and we get a bound

$$(3) \qquad |I_1| \leq \|g\|_1 \, AA^{-1} \ll (x + A^{-1})^{\varepsilon} \ll x^{\varepsilon/4}.$$

For the second integral I_2, again with $u = x - y$, we find that

$$|I_2| \leq \|g\|_\infty \int_{A^{-1}}^{A^{-1}x^{-1/8}} \frac{du}{u} \ll \|g\|_\infty \log 1/x$$

where the interval over which we take the sup norm of g is

$$A^{-1} \leq |x - y| < A^{-1}x^{-1/8}.$$

Using the growth estimate $g(x) = O(|x|^{\epsilon})$ for x near zero and the assumption $A^{-1} \leq x^{1/4}$, we get

$$\begin{aligned} |I_2| &\ll (x + A^{-1}x^{-1/8})^{\varepsilon} \log(1/x) \\ (4) \qquad &\ll (x + x^{1/8})^{\varepsilon} \log(1/x) \ll x^{\varepsilon/16}. \end{aligned}$$

For the third integral I_3, since the set of discontinuities of a function of bounded variation is countable, we can select x_0 such that g is continuous at x_0 and

$$x + A^{-1}x^{-1/8} \leq x_0 \leq x + 2A^{-1}x^{-1/8}.$$

For simplicity, we just look at $y \geq x_0$. The other piece for I_3 is done in the same way. We then decompose the integral over $y \geq x_0$ into a sum:

$$\begin{aligned} \int_{y \geq x_0} g(y) \frac{\sin A(x-y)}{x-y} dy &= \int_{y \geq x_0} g(x_0) \frac{\sin A(x-y)}{x-y} dy \\ &\quad + \int_{y \geq x_0} (g(y) - g(x_0)) \frac{\sin A(x-y)}{x-y} dy \\ &= J_1 + J_2, \end{aligned}$$

and we estimate J_1, J_2 successively. For the integral J_1, using Lemma 1.5, we find:

$$\begin{aligned} |J_1| &\ll |g(x_0)| \frac{1}{A(x_0 - x)} \\ &\ll (x + 2A^{-1}x^{-1/8})^{\varepsilon} A^{-1} A x^{1/8} \\ &\ll x^{1/8}. \end{aligned} \tag{5}$$

For the integral J_2, let χ_y be the characteristic function of $[0, y)$, that is put

$$\chi(y,t) = \chi_y(t) = \begin{cases} 1 & \text{if } t < y \\ 0 & \text{if } t \geq y. \end{cases}$$

If g is continuous at y, then

$$g(y) - g(x_0) = \int_{x_0}^{1} \chi(y,t) d\mu_g(t).$$

This is a convenient expression to plug into an application of Fubini's theorem which gives

$$\begin{aligned} J_2 &= \int_{x_0}^{1} \int_{x_0}^{1} \chi(y,t) \frac{\sin A(x-y)}{x-y} d\mu_g(t) dy \\ &= \int_{x_0}^{1} \left[\int_{x_0}^{1} \chi(y,t) \; \frac{\sin A(y-x)}{y-x} dy \right] d\mu_g(t) \\ &= \int_{x_0}^{1} \left[\int_{t}^{1} \frac{\sin A(y-x)}{y-x} \, dy \right] \; d\mu_g(t) \\ &= \int_{x_o}^{1} \left[\int_{t-x}^{1-x} \frac{\sin Au}{u} du \right] d\mu_g(t). \end{aligned}$$

Hence by Lemma 1.5, we find

$$\begin{aligned} |J_2| \leq \int_{x_0}^{1} \frac{3}{A(t-x)} |d\mu_g(t)| &\ll A^{-1}(x_0 - x)^{-1} V_{\mathbf{R}}(g) \\ &\ll A^{-1} A x^{1/8} = x^{1/8}, \end{aligned} \tag{6}$$

which concludes the estimate of the second integral J_2, and therefore of the integral I_3. Thus (3), (4), (5), and (6) conclude the proof of Lemma 1.9. □

Having taken care of all cases, we have completed the proof of Proposition 1.8.

To finish the proof of Theorem 1.6, we need the following proposition.

Proposition 1.10. *Let $g \in \mathrm{BV}(\mathbf{R}) \cap L^1(\mathbf{R})$. Then for all $A \geq 1$ we have*

$$\int_1^\infty g(y)\left[\frac{\sin Ay}{y} - \frac{\sin A(x-y)}{x-y}\right] dy = O_g(|x|^{1/4}) \text{ for } x \to 0,$$

the estimate on the right being independent of A.

The proof of Proposition 1.10, which is much easier than that of Proposition 1.8, will be given through the following lemmas.

Lemma 1.11. *Assume $g \in \mathrm{BV}(\mathbf{R}) \cap L^1(\mathbf{R})$. Then*

$$\left|\int_1^\infty g(y)\frac{\sin Ay}{y}dy\right| \leq \frac{3}{A}V_{\mathbf{R}}(g).$$

Proof. Extend g to $[0,\infty)$ by defining $g(t) = 0$ if $0 \leq t < 1$. For fixed $t \in (0,1)$ and $b > 1$, consider the integral

$$\int_t^b g(y)\frac{\sin Ay}{y}dy.$$

Let

$$S_A(x) = -\int_x^\infty \frac{\sin Ay}{y}dt$$

Using integration by parts we have

$$\int_t^b g(y)\frac{\sin Ay}{y}dy = g(y)S_A(y)\Big|_t^b - \int_t^b S_A(y)dg(y).$$

The point evaluations of $g(y)S_A(y)$ will be zero as b approaches infinity, since $g(t) = 0$, $g(b)$ approaches zero, and $S_A(y)$ is bounded. Also, by Lemma 1.5

$$\left| \int_t^b S_A(y)dg(y) \right| \leq \sup_{[t,b]} |S_A| \; V_{\mathbf{R}}(g) \leq \frac{3}{At} V_{\mathbf{R}}(g).$$

Therefore, we have shown, after letting t approach 1 and b approach ∞ that

$$\left| \int_1^\infty g(y) \frac{\sin Ay}{y} dy \right| \leq \frac{3}{A} V_{\mathbf{R}}(g),$$

as asserted. □

We now consider the integral in Proposition 1.10 in two separate cases, when $x \leq 1/A^4$ and when $x > 1/A^4$.

Case I. *Assume $x \leq 1/A^4$. Then there is a universal constant C such that*

$$\left| \int_1^\infty g(y) \left[\frac{\sin Ay}{y} - \frac{\sin A(x-y)}{x-y} \right] dy \right| \leq (C\|g\|_1)x^{1/2},$$

the estimate on the right independent of A.

Proof. By the Mean Value Theorem we have

$$\left| \frac{\sin A(y-x)}{y-x} - \frac{\sin Ay}{y} \right| \leq CA^2 x.$$

So, we can bound the integral in question by

$$CA^2 x\|g\|_1 \leq (C\|g\|_1)x^{1/2}. \quad \square$$

Case II. *Assume $x > 1/A^4$ and, without loss of generality, we can also assume that $x \leq 1/2$. Then*

$$\left| \int_1^\infty g(y) \left[\frac{\sin Ay}{y} - \frac{\sin A(x-y)}{x-y} \right] dy \right| \leq C_g x^{1/4},$$

where the constant C_g is independent of A, and depends only on g.

Proof. Let us estimate the integrals separately. In fact, write the integrals as

$$\int_1^\infty g(y) \frac{\sin A(y-x)}{y-x} dy - \int_1^\infty g(y) \frac{\sin Ay}{y} dy$$

$$= \int_{1-x}^\infty g(u+x) \frac{\sin Au}{u} du - \int_1^\infty g(y) \frac{\sin Ay}{y} dy$$

$$= \int_{1-x}^1 g(u+x) \frac{\sin Au}{u} du + \int_1^\infty g(u+x) \frac{\sin Au}{u} du$$

$$- \int_1^\infty g(y) \frac{\sin Ay}{y} dy.$$

The first integral is bounded by $\|g\|_\infty x$ since $(\sin Au)/u$ is bounded on $[1/2, 1]$. Lemma 1.11 applies to bound the second integral, independent of x, as well as the third integral. The bound achieved is

$$\|g\|_\infty x + \frac{6}{A} V_{\mathbf{R}}(g).$$

Since $1/A < x^{1/4}$, Case II is proved.

Combining the two cases above, we have completed our proof of Proposition 1.10. With all this, the proof of Theorem 1.6 is completed by combining Proposition 1.8 and Proposition 1.10.

§2. A Parseval formula

We recall the definition of the hermitian product

$$\langle f, g \rangle = \int_{-\infty}^{\infty} f(x)\bar{g}(x)dx.$$

For f, g in the Schwartz space, we assume the elementary **Parseval Formula**

$$\langle f, g \rangle = \langle f^{\wedge}, g^{\wedge} \rangle.$$

The point of this section is to prove this formula under less restrictive conditions. We shall extend conditions of Barner [Ba 90], for which the formula is true. The basic facts from real analysis (functions of bounded variation and Stieltjes integral, Fourier transforms under smooth conditions) used here are contained in [La 83].

We call the following conditions from [Ba 90] the **basic conditions** on a function f:

Condition 1. $f \in BV(\mathbf{R}) \cap L^1(\mathbf{R})$.

Condition 2. There exists $\varepsilon > 0$ such that

$$f(x) = f(0) + O(|x|^{\varepsilon}) \quad \text{for } x \to 0.$$

Condition 3. f is normalized, as defined in §1.

In addition to functions satisfying the basic conditions, we shall deal with a special pair of functions, arising as follows. We suppose given:

- A Borel measure μ on $\mathbf{R}^+$ such that $d\mu(x) = \psi(x)dx$, where ψ is some bounded, (Borel) measurable function.
- A measurable function φ on $\mathbf{R}^+$ such that:
  (a) The function $\varphi_0(x) = \varphi(x) - 1/x$ is bounded as x approaches zero.
  (b) Both functions $1/x$ and $\varphi(x)$ are in $L^1(|\mu|)$ outside a neighborhood of zero.

Thus we impose two asymptotic conditions on φ, one condition near zero and one condition at infinity. We call (μ, φ) a **special**

pair. We define the functional

$$W_{\mu,\varphi}(\alpha) = \int_0^\infty \left(\varphi(x)\alpha(x) - \frac{\alpha(0)}{x} \right) d\mu(x).$$

Its "Fourier transform" as a distribution is the function $W^\wedge_{\mu,\varphi}$ such that

$$\begin{aligned} W^\wedge_{\mu,\varphi}(t) &= \frac{1}{\sqrt{2\pi}} \int_0^\infty \left(\varphi(x) e^{-itx} - \frac{1}{x} \right) d\mu(x) \\ &= \frac{1}{\sqrt{2\pi}} W_{\mu,\varphi}(\bar{\chi}_t) \quad \text{where} \quad \chi_t(x) = e^{itx}. \end{aligned}$$

The following theorem generalizes the Barner-Weil formula [Ba 90].

Theorem 2.1. *Let f satisfy the three basic conditions, and let (μ, φ) be a special pair. Then*

$$\begin{aligned} \lim_{A \to \infty} \int_{-A}^{A} f^\wedge(t) W^\wedge_{\mu,\varphi}(t) dt &= W_{\mu,\varphi}(f^-) \\ &= \int_0^\infty \left[\varphi(x) f(-x) - \frac{f(0)}{x} \right] d\mu(x). \end{aligned}$$

Proof. At a certain point in the proof, we shall need to distinguish two cases, but we proceed as far as we can go without such distinction, according to a rather standard pattern of proof. We have:

$$\begin{aligned} &\int_{-A}^{A} f^\wedge(t) W^\wedge_{\mu,\varphi}(t) dt \\ &= \frac{1}{2\pi} \int_{-A}^{A} dt \int_{-\infty}^{\infty} f(y) e^{-ity} dy \int_0^\infty \left[\varphi(x) e^{-itx} - \frac{1}{x} \right] d\mu(x) \\ &= \frac{1}{2\pi} \int_{-A}^{A} dt \iint_{\mathbf{R} \times \mathbf{R}+} f(y) \left[\varphi(x) e^{-it(x+y)} - \frac{e^{-ity}}{x} \right] dy \, d\mu(x). \end{aligned}$$

By the assumptions on (μ, φ) and f, we can interchange the integrals which are absolutely convergent. We then perform the inner integration with respect to t, and the expression becomes

$$= \iint\limits_{\mathbf{R}\times\mathbf{R}+} \frac{f(y)}{\pi} \left[\varphi(x)\ \frac{\sin\ A(x+y)}{x+y} - \frac{\sin\ Ay}{xy}\right] dy d\mu(x)$$

$$= \int\limits_{\mathbf{R}+} \int\limits_{\mathbf{R}} \frac{f(-y)}{\pi} \left[\varphi(x)\frac{\sin A(x-y)}{x-y} - \frac{\sin Ay}{xy}\right] dy d\mu(x)$$

$$= \int\limits_{\mathbf{R}+} \left[\varphi(x) f_A(-x) - \frac{f_A(0)}{x}\right] d\mu(x).$$

At this point, we are finished with the proof in the case $f = \alpha$ is a function satisfying the conditions of Lemma 1.3, namely $\alpha_A = \alpha$ for sufficiently large A. For the general case, we write the above integral as a sum

$$= \int\limits_{\mathbf{R}+} (\varphi(x) - \frac{1}{x}) f_A(-x) d\mu(x) + \int\limits_{\mathbf{R}+} \frac{1}{x}(f_A(-x) - f_A(0)) d\mu(x).$$

Note that $\varphi(x) - 1/x$ is in $L^1(|\mu|)$ by our assumptions on (μ, φ).

Our final step is to prove that we can take the limit as $A \to \infty$ under the integral sign in both terms for an arbitrary f satisfying the basic conditions. In the first integral, one can apply the dominated convergence theorem by the boundedness of f_A (see Theorem 1.2) and the assumptions on $\varphi(x) - 1/x$. As to the second integral, we split the integral

$$\int\limits_{\mathbf{R}+} = \int_0^1 + \int_1^\infty .$$

Again, we apply the dominated convergence theory to the integral over $[1, \infty)$, by Theorem 1.2, the boundedness of f_A, and the assumption that $1/x$ is in $L^1(\mu)$. The more difficult part is the integral over the inteval $[0, 1]$.

Let α be as in Lemma 1.3 and such that $\alpha(0) = f(0)$, which can be achieved after multiplying α by a constant. Let $g = f - \alpha$. The formula of Theorem 2.1 is linear in f, and it is immediately verified that α and hence g satisfies the basic conditions and

$$g(x) = O(|x|^{\epsilon}) \quad \text{for } x \to 0.$$

Having proved the formula for α, we are reduced to proving it for g. Recall that Theorem 1.6 states that for a positive δ, which depends on ϵ,

$$g_A(x) - g_A(0) = O(|x|^{\delta}) \quad \text{for } x \to 0$$

uniformly in A. Therefore, the dominated convergence theorem again applies since $x^{-1+\delta}$ is integrable over $[0,1]$, and we get, since $g(0) = 0$,

$$\lim_{A\to\infty} \int_0^1 \frac{1}{x}(g_A(-x) - g_A(0))d\mu(x) = \int_0^1 \frac{1}{x} g(-x)d\mu(x).$$

After taking the limit as $A \to \infty$ under the integrals we see that the final expression above becomes

$$\int_{\mathbf{R}+} (\varphi(x) - \frac{1}{x})f(-x)d\mu(x) + \int_{\mathbf{R}+} \frac{1}{x}(f(-x) - f(0))d\mu(x),$$

which proves the theorem. □

§3. The General Parseval Formula

Following the ideas in [JoL 92a], we now prove a general Parseval formula associated to measurable functions with arbitrary principal part, thus generalizing the results of §2 in which the function $\varphi(x)$ was required to have principal part equal to $1/x$.

Suppose we are given:

- A Borel measure μ on $\mathbf{R}^+$ such that $d\mu(x) = \psi(x)dx$, where ψ is some bounded (Borel) measurable function.
- A measurable function φ on $\mathbf{R}^+$ having the following properties. There is a function $P_0(x) \in \mathbf{C}[\log x]\langle x \rangle$, which we shall write as

$$P_0(x) = \sum_{\mathrm{Re}(p)<0} b_p(x)x^p \quad \text{with} \quad b_p(x) = B_p(\log x) \in \mathbf{C}[\log x]$$

such that:

(a) There is some integer $m > 0$ such that

$$\varphi(x) - P_0(x) = O(|\log x|^m) \quad \text{for } x \to 0.$$

(b) Let M be the largest integer $< -\mathrm{Re}(p_0)$, so that

$$-1 \leq M + \mathrm{Re}(p_0) < 0.$$

Then both functions $x^M P_0(x)$ and $\varphi(x)$ are in $L^1(|\mu|)$ outside a neighborhood of zero.

From condition (a) and the power series expansion of e^{itx} one obtains the existence of functions $u_k(x)$ such that

$$\varphi(x)e^{itx} - \sum_{k=0}^{M} u_k(x)(it)^k = O(|\log x|^m) \quad \text{as } x \to 0.$$

The functions $u_k(x)$ come from the expression

$$\sum_{k+\mathrm{Re}(p)<0} \frac{b_p(x)x^{p+k}}{k!}(it)^k = \sum_{k=0}^{M} u_k(x)(it)^k.$$

As before, the above requirements impose two asymptotic conditions on φ, one condition near zero and one condition at infinity, and we call such a pair (μ, φ) a **special pair**. We define the functional

$$W_{\mu,\varphi}(\alpha) = \int_0^\infty \left(\varphi(x)\alpha(x) - \sum_{k=0}^{M} u_k(x)\alpha^{(k)}(0) \right) d\mu(x).$$

Its "Fourier transform" as a distribution is the function $W^\wedge_{\mu,\varphi}(t)$ such that

$$W^\wedge_{\mu,\varphi}(t) = \frac{1}{\sqrt{2\pi}} \int_0^\infty \left(\varphi(x)e^{-itx} - \sum_{k=0}^{M} u_k(x)(-it)^k \right) d\mu(x)$$

$$= \frac{1}{\sqrt{2\pi}} W_{\mu,\varphi}(\bar{\chi}_t) \quad \text{where} \quad \chi_t(x) = e^{itx}.$$

The following theorem generalizes Theorem 2.1.

Theorem 3.1. *Assume f and its first M derivatives satisfy the three basic conditions, and let (μ, φ) be a special pair. Then*

$$\lim_{A\to\infty} \int_{-A}^{A} f^\wedge(t) W^\wedge_{\mu,\varphi}(t)dt = W_{\mu,\varphi}(f^-)$$

$$= \int_0^\infty \left[\varphi(x)f(-x) - \sum_{k=0}^{M} u_k(x)(-1)^k f^{(k)}(0) \right] d\mu(x).$$

Proof. The proof is essentially identical to the proof of Theorem

2.1. For completeness, let us present the details. We have:

$$\int_{-A}^{A} f^{\wedge}(t)W^{\wedge}_{\mu,\varphi}(t)dt$$

$$= \int_{-A}^{A} \frac{dt}{2\pi} \int_{-\infty}^{\infty} f(y)e^{-ity}dy \int_{0}^{\infty} \left[\varphi(x)e^{-itx} - \sum_{k=0}^{M} u_k(x)(-it)^k\right] d\mu(x)$$

$$= \int_{-A}^{A} \frac{dt}{2\pi} \iint_{\mathbf{R}\times\mathbf{R}+} f(y)\left[\varphi(x)e^{-itx)} - \sum_{k=0}^{M} u_k(x)(-it)^k\right] e^{-ity}dyd\mu(x).$$

As in the proof of Theorem 2.1, the assumptions on (μ, φ) and f allow us to interchange the integrals which are absolutely convergent. By integrating with respect to t, this expression becomes

$$\iint_{\mathbf{R}\times\mathbf{R}+} \frac{f(y)}{\pi}\left[\varphi(x)\frac{\sin A(x+y)}{x+y} - \sum_{k=0}^{M} u_k(x)\left(\frac{d}{dy}\right)^k\left[\frac{\sin Ay}{y}\right]\right] dyd\mu(x).$$

Continuing, we have:

$$\iint_{\mathbf{R}\times\mathbf{R}+} \frac{f^{-}(y)}{\pi}\left[\varphi(x)\frac{\sin A(x-y)}{x-y} - \sum_{k=0}^{M} u_k(x)\left(\frac{-d}{dy}\right)^k\left[\frac{\sin Ay}{y}\right]\right] dyd\mu(x)$$

$$= \int_{\mathbf{R}+} \left[\varphi(x)f_A(-x) - \sum_{k=0}^{M} u_k(x)(-1)^k f_A^{(k)}(0)\right] d\mu(x).$$

In the above steps we have used the differentiation formula

$$\frac{1}{2}\int_{-A}^{A}(-it)^k e^{-ity}dt = \left(\frac{d}{dy}\right)^k\left[\frac{\sin Ay}{y}\right]$$

and the integration by parts formula

$$\int_{-\infty}^{\infty}\left[\left(-\frac{d}{dy}\right)^k\frac{\sin Ay}{y}\right]\frac{f(-y)}{\pi}dy = (-1)^k f_A^{(k)}(0),$$

which is valid by Lemma 1.1 and the arguments given in the proof of Corollary 1.7.

At this point, we are finished with the proof in the case $f = \alpha$ is a function satisfying the conditions of Lemma 1.3, namely $\alpha_A = \alpha$ for sufficiently large A. For the general case, we write the integral as the sum

$$\int_{\mathbf{R}+} f_A(-x)(\varphi(x) - P_0(x))d\mu(x)$$

$$+ \int_1^\infty \left[f_A(-x)P_0(x) - \sum_{k=0}^{M} u_k(x)(-1)^k f_A^{(k)}(0) \right] d\mu(x)$$

$$+ \int_0^1 \left[f_A(-x)P_0(x) - \sum_{k=0}^{M} u_k(x)(-1)^k f_A^{(k)}(0) \right] d\mu(x).$$

The proof of Theorem 3.1 now finishes as did the proof of Theorem 2.1. By an appropriate extension of Lemma 1.3, choose an α for which $\alpha_A = \alpha$ for sufficiently large A and the numbers $\alpha(0), \ldots, \alpha^{(M)}(0)$ have been chosen to agree with the first M derivatives of f at zero, and set $g = f - \alpha$. The above integrals are linear in the function f, so proving the theorem for g will imply the theorem for f, so we work with g. By Theorem 1.2, the boundedness of g_A (as stated in Lemma 1.1), and assumption (b) above, one can apply the dominated convergence theorem to the first two integrals above. Using Corollary 1.7 and the definition of the function $u_k(x)$, we can write the third integral as

$$\int_0^1 \left[g_A(-x)P_0(x) - \sum_{k=0}^{M} u_k(x)(-1)^k g_A^{(k)}(0) \right] d\mu(x)$$

$$= \int_0^1 P_0(x) \left[g_A(-x) - \sum_{k=0}^{M} \frac{g_A^{(k)}(0)}{k!}(-x)^k \right] d\mu(x).$$

By Corollary 1.7, the integrand is bounded by $Cx^{M+\mathrm{Re}(p_0)+\delta}$ and

$$M + \mathrm{Re}(p_0) + \delta \geq -1 + \delta,$$

so $Cx^{M+\mathrm{Re}(p_0)+\delta}$ is integrable. The dominated convergence theorem applies, and the theorem is proved. $\square$

§4. The Parseval Formula for $I_w(a+it)$

To conclude our investigation of the regularized harmonic series, let us show how Theorem 3.1 applies to prove a Parseval formula associated to the regularized harmonic series encountered in §3 of [JoL 92a]. We will assume the notation defined in §3.

Recall as in §4 of [JoL 92a] that the classical Gauss formula states that for $\mathrm{Re}(z) > 0$,

$$-\Gamma'/\Gamma(z+1) = \int_0^\infty \left[\frac{e^{-zx}}{1-e^{-x}} - \frac{1}{x}\right] e^{-x}\,dx.$$

If we let $z = a + it$ with $a > -1$, then we get

$$-\Gamma'/\Gamma(a+1+it) = \int_0^\infty \left[\varphi_a(x)e^{-itx} - \frac{1}{x}\right] d\mu(x) \tag{1}$$

where

$$\varphi_a(x) = \frac{e^{-ax}}{1-e^{-x}} \quad \text{and} \quad d\mu(x) = e^{-x}\,dx.$$

One can view (1) as a type of regularized Fourier transform representation of the gamma function. Finally, by the change of variables t to t/b then x to bx for $b > 0$, (1) becomes

$$-\Gamma'/\Gamma(a+1+i\frac{t}{b}) = \int_0^\infty \left[\frac{be^{-abx}}{1-e^{-bx}}e^{-itx} - \frac{1}{x}\right] e^{-bx}\,dx.$$

Next we will present a generalization of (1) making use of results in §5 of [JoL 92a]. That is, associated to any Dirichlet series ζ satisfying **DIR 1** and **DIR 2**, as defined in §1 of [JoL 92a], and whose associated theta function $\theta(t)$ satisfies **AS 1**, **AS 2** and **AS 3**, we will realize the regularized harmonic series $I_w(z)$, whose definition will be recalled below, as a regularized Fourier transform. Theorem 2.1 applies to the classical Parseval formula involving the gamma function, which is used in the Barner-Weil explicit formula. In this section we will use our regularized Fourier transform to present a general Parseval formula.

Recall that the principal part of the theta function is

$$P_0\theta(x) = \sum_{\mathrm{Re}(p)<0} b_p(x)x^p,$$

so

$$\theta(x) - P_0\theta(x) = O(|\log x|^m) \quad \text{as } x \to 0.$$

As in [JoL 92a], let $\theta_z(x) = e^{-zx}\theta(x)$. By expanding e^{-zx} in a power series, we see that the principal part of $\theta_z(x)$ is

$$\text{(2)} \quad P_0\theta_z(x) = P_0\left[e^{-zx}\theta(x)\right] = \sum_{\mathrm{Re}(p)+k<0} \frac{b_p(x)x^{p+k}}{k!}(-z)^k.$$

From §4 of [JoL 92a] we recall the following result.

Theorem 4.1. *For any fixed complex* w *with*

$$\mathrm{Re}(w) > \max_k\{-\mathrm{Re}(\lambda_k)\} \quad \text{and} \quad \mathrm{Re}(w) > 0,$$

the integral

$$I_w(z) = \int_0^\infty [\theta_z(x) - P_0\theta_z(x)]\, e^{-wx}dx$$

is convergent for $\mathrm{Re}(z) > 0$. *Further,* $I_w(z)$ *has a meromorphic continuation to all* $z \in \mathbf{C}$ *with simple poles at* $-\lambda_k + w$ *with residue* a_k.

Remark. In the spectral case, when $\zeta(s) = \sum a_k\lambda_k^{-s}$ with $a_k \in \mathbf{Z}_{\geq 0}$, Theorem 4.1 of [JoL 92a] states that

$$D_L'/D_L(z+w) = I_w(z) + S_w(z)$$

where $S_w(z)$ is a polynomial in z of degree $< -\mathrm{Re}(p_0)$, with coefficients whose dependence on L is through b_p for $\mathrm{Re}(p) < 0$, and whose dependence on w is through elements in $\mathbf{C}[\log w]\langle w\rangle$. Also, in §4 of [JoL 92a] it is shown that Theorem 4.1 yields the classical Gauss formula in the case $L = \mathbf{Z}_{\geq 0}$.

To continue, let us work with the principal part of the theta function. If we restrict the variable z in (2) to a vertical line by letting $z = a + it$ we get

$$P_0\theta_z(x) = \sum_{\mathrm{Re}(p)+k<0} \frac{b_p(x)x^{p+k}}{k!}(-a-it)^k$$

$$(3) \qquad = \sum_{k<-\mathrm{Re}(p_0)} c_k(a,x)(-it)^k,$$

where the coefficients

$$c_k(a,x) = c_k(a,x,\zeta)$$

depend on the variables a, x and on ζ through the coefficients of t^p for $\mathrm{Re}(p) < 0$ (see **AS 2**). With this, the integral in Theorem 4.1 can be written as

(4)

$$I_w(z) = \int_0^\infty \left[\theta(x)e^{-x(a+it)} - \sum_{k<-\mathrm{Re}(p_0)} c_k(a,x)(-it)^k\right] e^{-wx}\,dx.$$

Thus, we obtain:

Corollary 4.2. *For any $w \in \mathbf{C}$ with $\mathrm{Re}(w) > \max_k\{-\mathrm{Re}(\lambda_k)\}$ and $\mathrm{Re}(w) > 0$, and any $a \in \mathbf{R}^+$ define*

$$d\mu_w(x) = e^{-wx}dx \quad \text{and} \quad \theta_a(x) = \theta(x)e^{-ax}.$$

Then

$$I_w(a+it) = \int_0^\infty \left[\theta_a(x)e^{-itx} - \sum_{k<-\mathrm{Re}(p_0)} c_k(a,x)(-it)^k\right] d\mu_w(x).$$

This is our desired generalization of (1) and will be referred to as the **regularized Fourier transform** representation of a regularized harmonic series. Then Theorem 3.1 yields:

Theorem 4.3. *Assume f and its first M derivatives satisfy the three basic conditions. For any $w \in \mathbf{C}$ with*

$$\mathrm{Re}(w) > \max_k\{-\mathrm{Re}(\lambda_k)\} \quad \text{and} \quad \mathrm{Re}(w) > 0,$$

and any $a \in \mathbf{R}^+$, define

$$d\mu_w(x) = e^{-wx}dx \quad \text{and} \quad \theta_a(x) = \theta(x)e^{-ax}.$$

Then

$$\lim_{A\to\infty} \frac{1}{\sqrt{2\pi}} \int_{-A}^{A} f^\wedge(t) I_w(a+it)dt$$

$$= \int_0^\infty \left[\theta_a(x)f(-x) - \sum_{k<-\mathrm{Re}(p_0)} c_k(a,x)f^{(k)}(0) \right] d\mu_w(x).$$

In the spectral case when $L = \mathbf{Z}_{\geq 0}$ Theorem 4.3 is the classical Barner-Weil formula.

BIBLIOGRAPHY

[Ba 81] BARNER, K.: On Weil's explicit formula. *J. reine angew. Math.* **323,** 139-152 (1981).

[Ba 90] BARNER, K.: Einführung in die Analytische Zahlentheorie. Preprint (1990).

[BeKn 86] BELAVIN, A. A., and KNIZHNIK, V. G.: Complex geometry and the theory of quantum string. *Sov. Phy. JETP* **2,** 214-228 (1986).

[BrS 85] BRÜNING, J., and SEELEY, R.: Regular singular asymptotics. *Adv. Math.* **58,** 133-148 (1985).

[CaV 90] CARTIER, P., and VOROS, A.: Une nouvelle interprétation de la formule des traces de Selberg. pp. 1-68, volume 87 of *Progress in Mathematics* , Boston: Birkhauser (1990).

[Cr 19] CRAMÉR, H.: Studien über die Nullstellen der Riemannschen Zetafunktion. *Math. Z.* **4,** 104-130 (1919).

[De 92] DENINGER, C.: Local L-factors of motives and regularized products. *Invent. Math.* **107,** 135-150 (1992).

[DP 86] D'HOKER, E., and PHONG, D.: On determinants of Laplacians on Riemann surfaces. *Comm. Math. Phys.* **105,** 537-545 (1986).

[Di 78] DIEUDONNÉ, J.: *Éléments d' Analyse, Vol. VII,* Paris: Gauthier-Villars (1978), reprinted as *Treatise on Analysis Volume 10-VII* San Diego: Academic Press (1988).

[DG 75] DUISTERMAAT, J. J. and GUILLEMIN, V. W.: The spectrum of positive elliptic operators and periodic bicharacteristics. *Invent. Math.* **29,** 39-79 (1975).

[Fr 86] FRIED, D.: Analytic torsion and closed geodesics on hyperbolic manifolds. *Invent. Math.* **84,** 523-540 (1986).

[G 77] GANGOLLI, R.: Zeta functions of Selberg's type for compact space forms of symmetric space of rank one. *Ill. J. Math.* **21,** 1-42 (1977).

[Gr 86] GRUBB, G.: *Functional Calculus of Pseudo-Differential Boundary Problems.* Progress in Mathematics **65** Boston: Birkhauser (1986).

[Ha 49] HARDY, G. H.: *Divergent Series.* Oxford: Oxford University Press (1949).

[In 32] INGHAM, A.: *The Distribution of Prime Numbers,* Cambridge University Press, Cambridge, (1932).

[JoL 92a] JORGENSON, J., and LANG, S.: Some complex analytic properties of regularized products and series. This volume.

[JoL 92b] JORGENSON, J., and LANG, S.: On Cramér's theorem for general Euler products with functional equations. To appear in *Math. Annalen.*

[JoL 92c] JORGENSON, J., and LANG, S.: A Parseval formula for functions with a singular asymptotic expansion at the origin. This volume.

[JoL 92d] JORGENSON, J., and LANG, S.: Artin formalism and heat kernels. To appear in *J. reine angew. Math.*

[JoL 93] JORGENSON, J., and LANG, S.: Explicit formulas and regularized products. Yale University Preprint (1993).

[Ku 88] KUROKAWA, N.: Parabolic components of zeta functions. *Proc. Japan Acad., Ser A* **64,** 21-24 (1988).

[La 70] LANG, S.: *Algebraic Number Theory,* Menlo Park: Addison-Wesley (1970), reprinted as Graduate Texts in Mathematics **110,** New York: Springer-Verlag (1986).

[La 83] LANG, S.: *Real Analysis,* Menlo Park: Addison-Wesley (1983).

[La 85] LANG, S.: *Complex Analysis,* Graduate Texts in Mathematics **103,** New York: Springer-Verlag (1985).

[La 93] LANG, S.: *Real and Functional Analysis, Third Edition*, New York: Springer-Verlag (1993).

[McS 67] McKEAN, H. P., and SINGER, I. M.: Curvature and eigenvalues of the Laplacian. *J. Differ. Geom.* **1,** 43-70 (1967).

[MP 49] MINAKSHISUNDARAM, S., and PLEIJEL, A.: Some properties of the eigenfunctions of the Laplace operator on Riemannian manifolds. *Can. J. Math.* **1,** 242-256 (1949).

[RS 73] RAY, D., and SINGER, I.: Analytic torsion for complex manifolds. *Ann. Math.* **98,** 154-177 (1973).

[Sa 87] SARNAK, P.: Determinants of Laplacians. *Comm. Math. Phys.* **110,** 113-120 (1987).

[Su 85] SUNADA, T.: Riemannian coverings and isospectral manifolds. *Ann. Math.* **121,** 169-186 (1985).

[TaZ 91] TAKHTAJAN, L. A., and ZOGRAF, P. G.: A local index theorem for families of $\bar{\partial}$-operators on punctured Riemann surfaces and a new Kähler metric on their moduli spaces. *Comm. Math. Phys.* **137,** 399-426 (1991).

[Ti 48] TITCHMARSH, E. C.: *Introduction to the Theory of Fourier-Integrals, 2nd Edition* Oxford University Press, Oxford (1948).

[Ve 81] VENKOV, A. B., and ZOGRAF, P. G.: On analogues of the Artin factorization formulas in the spectral theory of automorphic functions connected with induced representations of Fuschsian groups, *Soviet Math. Dokl.,* **21,** 94-96 (1981)

[Ve 83] VENKOV, A. B., and ZOGRAF, P. G.: On analogues of the Artin factorization formulas in the spectral theory of automorphic functions connected with induced representations of Fuschsian groups, *Math. USSR- Izv.* **21,** 435-443 (1983).

[Vo 87] VOROS, A.: Spectral functions, special functions, and the Selberg zeta function. *Comm. Math. Phys* **110,** 439-465 (1987).

[We 52] WEIL, A.: Sur les "formules explicites" de la théorie des nombres premiers, *Comm. Lund* (vol. dédié à Marcel Riesz), 252-265 (1952).

[We 72] WEIL, A.: Sur les formules explicites de la théorie des nombres, *Izv. Mat. Nauk (Ser. Mat.)* **36,** 3-18 (1972).

Printing: Weihert-Druck GmbH, Darmstadt
Binding: Buchbinderei Schäffer, Grünstadt

Lecture Notes in Mathematics

For information about Vols. 1–1384
please contact your bookseller or Springer-Verlag

Vol. 1425: R.A. Piccinini (Ed.), Groups of Self-Equivalences and Related Topics. Proceedings, 1988. V, 214 pages. 1990.

Vol. 1426: J. Azéma, P.A. Meyer, M. Yor (Eds.), Séminaire de Probabilités XXIV, 1988/89. V, 490 pages. 1990.

Vol. 1427: A. Ancona, D. Geman, N. Ikeda, École d'Eté de Probabilités de Saint Flour XVIII, 1988. Ed.: P.L. Hennequin. VII, 330 pages. 1990.

Vol. 1428: K. Erdmann, Blocks of Tame Representation Type and Related Algebras. XV. 312 pages. 1990.

Vol. 1429: S. Homer, A. Nerode, R.A. Platek, G.E. Sacks, A. Scedrov, Logic and Computer Science. Seminar, 1988. Editor: P. Odifreddi. V, 162 pages. 1990.

Vol. 1430: W. Bruns, A. Simis (Eds.), Commutative Algebra. Proceedings. 1988. V, 160 pages. 1990.

Vol. 1431: J.G. Heywood, K. Masuda, R. Rautmann, V.A. Solonnikov (Eds.), The Navier-Stokes Equations – Theory and Numerical Methods. Proceedings, 1988. VII, 238 pages. 1990.

Vol. 1432: K. Ambos-Spies, G.H. Müller, G.E. Sacks (Eds.), Recursion Theory Week. Proceedings, 1989. VI, 393 pages. 1990.

Vol. 1433: S. Lang, W. Cherry, Topics in Nevanlinna Theory. II, 174 pages.1990.

Vol. 1434: K. Nagasaka, E. Fouvry (Eds.), Analytic Number Theory. Proceedings, 1988. VI, 218 pages. 1990.

Vol. 1435: St. Ruscheweyh, E.B. Saff, L.C. Salinas, R.S. Varga (Eds.), Computational Methods and Function Theory. Proceedings, 1989. VI, 211 pages. 1990.

Vol. 1436: S. Xambó-Descamps (Ed.), Enumerative Geometry. Proceedings, 1987. V, 303 pages. 1990.

Vol. 1437: H. Inassaridze (Ed.), K-theory and Homological Algebra. Seminar, 1987–88. V, 313 pages. 1990.

Vol. 1438: P.G. Lemarié (Ed.) Les Ondelettes en 1989. Seminar. IV, 212 pages. 1990.

Vol. 1439: E. Bujalance, J.J. Etayo, J.M. Gamboa, G. Gromadzki. Automorphism Groups of Compact Bordered Klein Surfaces: A Combinatorial Approach. XIII, 201 pages. 1990.

Vol. 1440: P. Latiolais (Ed.), Topology and Combinatorial Groups Theory. Seminar, 1985–1988. VI, 207 pages. 1990.

Vol. 1441: M. Coornaert, T. Delzant, A. Papadopoulos. Géométrie et théorie des groupes. X, 165 pages. 1990.

Vol. 1442: L. Accardi, M. von Waldenfels (Eds.), Quantum Probability and Applications V. Proceedings, 1988. VI, 413 pages. 1990.

Vol. 1443: K.H. Dovermann, R. Schultz, Equivariant Surgery Theories and Their Periodicity Properties. VI, 227 pages. 1990.

Vol. 1444: H. Korezlioglu, A.S. Ustunel (Eds.), Stochastic Analysis and Related Topics VI. Proceedings, 1988. V, 268 pages. 1990.

Vol. 1445: F. Schulz, Regularity Theory for Quasilinear Elliptic Systems and – Monge Ampère Equations in Two Dimensions. XV, 123 pages. 1990.

Vol. 1446: Methods of Nonconvex Analysis. Seminar, 1989. Editor: A. Cellina. V, 206 pages. 1990.

Vol. 1447: J.-G. Labesse, J. Schwermer (Eds), Cohomology of Arithmetic Groups and Automorphic Forms. Proceedings, 1989. V, 358 pages. 1990.

Vol. 1448: S.K. Jain, S.R. López-Permouth (Eds.), Non-Commutative Ring Theory. Proceedings, 1989. V, 166 pages. 1990.

Vol. 1449: W. Odyniec, G. Lewicki, Minimal Projections in Banach Spaces. VIII, 168 pages. 1990.

Vol. 1450: H. Fujita, T. Ikebe, S.T. Kuroda (Eds.), Functional-Analytic Methods for Partial Differential Equations. Proceedings, 1989. VII, 252 pages. 1990.

Vol. 1451: L. Alvarez-Gaumé, E. Arbarello, C. De Concini, N.J. Hitchin, Global Geometry and Mathematical Physics. Montecatini Terme 1988. Seminar. Editors: M. Francaviglia, F. Gherardelli. IX, 197 pages. 1990.

Vol. 1452: E. Hlawka, R.F. Tichy (Eds.), Number-Theoretic Analysis. Seminar, 1988–89. V, 220 pages. 1990.

Vol. 1453: Yu.G. Borisovich, Yu.E. Gliklikh (Eds.), Global Analysis – Studies and Applications IV. V, 320 pages. 1990.

Vol. 1454: F. Baldassari, S. Bosch, B. Dwork (Eds.), p-adic Analysis. Proceedings, 1989. V, 382 pages. 1990.

Vol. 1455: J.-P. Françoise, R. Roussarie (Eds.), Bifurcations of Planar Vector Fields. Proceedings, 1989. VI, 396 pages. 1990.

Vol. 1456: L.G. Kovács (Ed.), Groups – Canberra 1989. Proceedings. XII, 198 pages. 1990.

Vol. 1457: O. Axelsson, L.Yu. Kolotilina (Eds.), Preconditioned Conjugate Gradient Methods. Proceedings, 1989. V, 196 pages. 1990.

Vol. 1458: R. Schaaf, Global Solution Branches of Two Point Boundary Value Problems. XIX, 141 pages. 1990.

Vol. 1459: D. Tiba, Optimal Control of Nonsmooth Distributed Parameter Systems. VII, 159 pages. 1990.

Vol. 1460: G. Toscani, V. Boffi, S. Rionero (Eds.), Mathematical Aspects of Fluid Plasma Dynamics. Proceedings, 1988. V, 221 pages. 1991.

Vol. 1461: R. Gorenflo, S. Vessella, Abel Integral Equations. VII, 215 pages. 1991.

Vol. 1462: D. Mond, J. Montaldi (Eds.), Singularity Theory and its Applications. Warwick 1989, Part I. VIII, 405 pages. 1991.

Vol. 1463: R. Roberts, I. Stewart (Eds.), Singularity Theory and its Applications. Warwick 1989, Part II. VIII, 322 pages. 1991.

Vol. 1464: D. L. Burkholder, E. Pardoux, A. Sznitman, Ecole d'Eté de Probabilités de Saint- Flour XIX-1989. Editor: P. L. Hennequin. VI, 256 pages. 1991.

Vol. 1465: G. David, Wavelets and Singular Integrals on Curves and Surfaces. X, 107 pages. 1991.

Vol. 1466: W. Banaszczyk, Additive Subgroups of Topological Vector Spaces. VII, 178 pages. 1991.

Vol. 1467: W. M. Schmidt, Diophantine Approximations and Diophantine Equations. VIII, 217 pages. 1991.

Vol. 1468: J. Noguchi, T. Ohsawa (Eds.), Prospects in Complex Geometry. Proceedings, 1989. VII, 421 pages. 1991.

Vol. 1469: J. Lindenstrauss, V. D. Milman (Eds.), Geometric Aspects of Functional Analysis. Seminar 1989-90. XI, 191 pages. 1991.

Vol. 1470: E. Odell, H. Rosenthal (Eds.), Functional Analysis. Proceedings, 1987-89. VII, 199 pages. 1991.

Vol. 1471: A. A. Panchishkin, Non-Archimedean L-Functions of Siegel and Hilbert Modular Forms. VII, 157 pages. 1991.

Vol. 1472: T. T. Nielsen, Bose Algebras: The Complex and Real Wave Representations. V, 132 pages. 1991.

Vol. 1473: Y. Hino, S. Murakami, T. Naito, Functional Differential Equations with Infinite Delay. X, 317 pages. 1991.

Vol. 1474: S. Jackowski, B. Oliver, K. Pawałowski (Eds.), Algebraic Topology, Poznań 1989. Proceedings. VIII, 397 pages. 1991.

Vol. 1475: S. Busenberg, M. Martelli (Eds.), Delay Differential Equations and Dynamical Systems. Proceedings, 1990. VIII, 249 pages. 1991.

Vol. 1476: M. Bekkali, Topics in Set Theory. VII, 120 pages. 1991.

Vol. 1477: R. Jajte, Strong Limit Theorems in Noncommutative L_2-Spaces. X, 113 pages. 1991.

Vol. 1478: M.-P. Malliavin (Ed.), Topics in Invariant Theory. Seminar 1989-1990. VI, 272 pages. 1991.

Vol. 1479: S. Bloch, I. Dolgachev, W. Fulton (Eds.), Algebraic Geometry. Proceedings, 1989. VII, 300 pages. 1991.

Vol. 1480: F. Dumortier, R. Roussarie, J. Sotomayor, H. Żoładek, Bifurcations of Planar Vector Fields: Nilpotent Singularities and Abelian Integrals. VIII, 226 pages. 1991.

Vol. 1481: D. Ferus, U. Pinkall, U. Simon, B. Wegner (Eds.), Global Differential Geometry and Global Analysis. Proceedings, 1991. VIII, 283 pages. 1991.

Vol. 1482: J. Chabrowski, The Dirichlet Problem with L^2-Boundary Data for Elliptic Linear Equations. VI, 173 pages. 1991.

Vol. 1483: E. Reithmeier, Periodic Solutions of Nonlinear Dynamical Systems. VI, 171 pages. 1991.

Vol. 1484: H. Delfs, Homology of Locally Semialgebraic Spaces. IX, 136 pages. 1991.

Vol. 1485: J. Azéma, P. A. Meyer, M. Yor (Eds.), Séminaire de Probabilités XXV. VIII, 440 pages. 1991.

Vol. 1486: L. Arnold, H. Crauel, J.-P. Eckmann (Eds.), Lyapunov Exponents. Proceedings, 1990. VIII, 365 pages. 1991.

Vol. 1487: E. Freitag, Singular Modular Forms and Theta Relations. VI, 172 pages. 1991.

Vol. 1488: A. Carboni, M. C. Pedicchio, G. Rosolini (Eds.), Category Theory. Proceedings, 1990. VII, 494 pages. 1991.

Vol. 1489: A. Mielke, Hamiltonian and Lagrangian Flows on Center Manifolds. X, 140 pages. 1991.

Vol. 1490: K. Metsch, Linear Spaces with Few Lines. XIII, 196 pages. 1991.

Vol. 1491: E. Lluis-Puebla, J.-L. Loday, H. Gillet, C. Soulé, V. Snaith, Higher Algebraic K-Theory: an overview. IX, 164 pages. 1992.

Vol. 1492: K. R. Wicks, Fractals and Hyperspaces. VIII, 168 pages. 1991.

Vol. 1493: E. Benoît (Ed.), Dynamic Bifurcations. Proceedings, Luminy 1990. VII, 219 pages. 1991.

Vol. 1494: M.-T. Cheng, X.-W. Zhou, D.-G. Deng (Eds.), Harmonic Analysis. Proceedings, 1988. IX, 226 pages. 1991.

Vol. 1495: J. M. Bony, G. Grubb, L. Hörmander, H. Komatsu, J. Sjöstrand, Microlocal Analysis and Applications. Montecatini Terme, 1989. Editors: L. Cattabriga, L. Rodino. VII, 349 pages. 1991.

Vol. 1496: C. Foias, B. Francis, J. W. Helton, H. Kwakernaak, J. B. Pearson, H_∞-Control Theory. Como, 1990. Editors: E. Mosca, L. Pandolfi. VII, 336 pages. 1991.

Vol. 1497: G. T. Herman, A. K. Louis, F. Natterer (Eds.), Mathematical Methods in Tomography. Proceedings 1990. X, 268 pages. 1991.

Vol. 1498: R. Lang, Spectral Theory of Random Schrödinger Operators. X, 125 pages. 1991.

Vol. 1499: K. Taira, Boundary Value Problems and Markov Processes. IX, 132 pages. 1991.

Vol. 1500: J.-P. Serre, Lie Algebras and Lie Groups. VII, 168 pages. 1992.

Vol. 1501: A. De Masi, E. Presutti, Mathematical Methods for Hydrodynamic Limits. IX, 196 pages. 1991.

Vol. 1502: C. Simpson, Asymptotic Behavior of Monodromy. V, 139 pages. 1991.

Vol. 1503: S. Shokranian, The Selberg-Arthur Trace Formula (Lectures by J. Arthur). VII, 97 pages. 1991.

Vol. 1504: J. Cheeger, M. Gromov, C. Okonek, P. Pansu, Geometric Topology: Recent Developments. Editors: P. de Bartolomeis, F. Tricerri. VII, 197 pages. 1991.

Vol. 1505: K. Kajitani, T. Nishitani, The Hyperbolic Cauchy Problem. VII, 168 pages. 1991.

Vol. 1506: A. Buium, Differential Algebraic Groups of Finite Dimension. XV, 145 pages. 1992.

Vol. 1507: K. Hulek, T. Peternell, M. Schneider, F.-O. Schreyer (Eds.), Complex Algebraic Varieties. Proceedings, 1990. VII, 179 pages. 1992.

Vol. 1508: M. Vuorinen (Ed.), Quasiconformal Space Mappings. A Collection of Surveys 1960-1990. IX, 148 pages. 1992.

Vol. 1509: J. Aguadé, M. Castellet, F. R. Cohen (Eds.), Algebraic Topology - Homotopy and Group Cohomology. Proceedings, 1990. X, 330 pages. 1992.

Vol. 1510: P. P. Kulish (Ed.), Quantum Groups. Proceedings, 1990. XII, 398 pages. 1992.

Vol. 1511: B. S. Yadav, D. Singh (Eds.), Functional Analysis and Operator Theory. Proceedings, 1990. VIII, 223 pages. 1992.

Vol. 1512: L. M. Adleman, M.-D. A. Huang, Primality Testing and Abelian Varieties Over Finite Fields. VII, 142 pages. 1992.

Vol. 1513: L. S. Block, W. A. Coppel, Dynamics in One Dimension. VIII, 249 pages. 1992.

Vol. 1514: U. Krengel, K. Richter, V. Warstat (Eds.), Ergodic Theory and Related Topics III, Proceedings, 1990. VIII, 236 pages. 1992.

Vol. 1515: E. Ballico, F. Catanese, C. Ciliberto (Eds.), Classification of Irregular Varieties. Proceedings, 1990. VII, 149 pages. 1992.

Vol. 1516: R. A. Lorentz, Multivariate Birkhoff Interpolation. IX, 192 pages. 1992.

Vol. 1517: K. Keimel, W. Roth, Ordered Cones and Approximation. VI, 134 pages. 1992.

Vol. 1518: H. Stichtenoth, M. A. Tsfasman (Eds.), Coding Theory and Algebraic Geometry. Proceedings, 1991. VIII, 223 pages. 1992.

Vol. 1519: M. W. Short, The Primitive Soluble Permutation Groups of Degree less than 256. IX, 145 pages. 1992.

Vol. 1520: Yu. G. Borisovich, Yu. E. Gliklikh (Eds.), Global Analysis – Studies and Applications V. VII, 284 pages. 1992.

Vol. 1521: S. Busenberg, B. Forte, H. K. Kuiken, Mathematical Modelling of Industrial Process. Bari, 1990. Editors: V. Capasso, A. Fasano. VII, 162 pages. 1992.

Vol. 1522: J.-M. Delort, F. B. I. Transformation. VII, 101 pages. 1992.

Vol. 1523: W. Xue, Rings with Morita Duality. X, 168 pages. 1992.

Vol. 1524: M. Coste, L. Mahé, M.-F. Roy (Eds.), Real Algebraic Geometry. Proceedings, 1991. VIII, 418 pages. 1992.

Vol. 1525: C. Casacuberta, M. Castellet (Eds.), Mathematical Research Today and Tomorrow. VII, 112 pages. 1992.

Vol. 1526: J. Azéma, P. A. Meyer, M. Yor (Eds.), Séminaire de Probabilités XXVI. X, 633 pages. 1992.

Vol. 1527: M. I. Freidlin, J.-F. Le Gall, Ecole d'Eté de Probabilités de Saint-Flour XX – 1990. Editor: P. L. Hennequin. VIII, 244 pages. 1992.

Vol. 1528: G. Isac, Complementarity Problems. VI, 297 pages. 1992.

Vol. 1529: J. van Neerven, The Adjoint of a Semigroup of Linear Operators. X, 195 pages. 1992.

Vol. 1530: J. G. Heywood, K. Masuda, R. Rautmann, S. A. Solonnikov (Eds.), The Navier-Stokes Equations II – Theory and Numerical Methods. IX, 322 pages. 1992.

Vol. 1531: M. Stoer, Design of Survivable Networks. IV, 206 pages. 1992.

Vol. 1532: J. F. Colombeau, Multiplication of Distributions. X, 184 pages. 1992.

Vol. 1533: P. Jipsen, H. Rose, Varieties of Lattices. X, 162 pages. 1992.

Vol. 1534: C. Greither, Cyclic Galois Extensions of Commutative Rings. X, 145 pages. 1992.

Vol. 1535: A. B. Evans, Orthomorphism Graphs of Groups. VIII, 114 pages. 1992.

Vol. 1536: M. K. Kwong, A. Zettl, Norm Inequalities for Derivatives and Differences. VII, 150 pages. 1992.

Vol. 1537: P. Fitzpatrick, M. Martelli, J. Mawhin, R. Nussbaum, Topological Methods for Ordinary Differential Equations. Montecatini Terme, 1991. Editors: M. Furi, P. Zecca. VII, 218 pages. 1993.

Vol. 1538: P.-A. Meyer, Quantum Probability for Probabilists. X, 287 pages. 1993.

Vol. 1539: M. Coornaert, A. Papadopoulos, Symbolic Dynamics and Hyperbolic Groups. VIII, 138 pages. 1993.

Vol. 1540: H. Komatsu (Ed.), Functional Analysis and Related Topics, 1991. Proceedings. XXI, 413 pages. 1993.

Vol. 1541: D. A. Dawson, B. Maisonneuve, J. Spencer, Ecole d' Eté de Probabilités de Saint-Flour XXI - 1991. Editor: P. L. Hennequin. VIII, 356 pages. 1993.

Vol. 1542: J.Fröhlich, Th.Kerler, Quantum Groups, Quantum Categories and Quantum Field Theory. VII, 431 pages. 1993.

Vol. 1543: A. L. Dontchev, T. Zolezzi, Well-Posed Optimization Problems. XII, 421 pages. 1993.

Vol. 1544: M.Schürmann, White Noise on Bialgebras. VII, 146 pages. 1993.

Vol. 1545: J. Morgan, K. O'Grady, Differential Topology of Complex Surfaces. VIII, 224 pages. 1993.

Vol. 1546: V. V. Kalashnikov, V. M. Zolotarev (Eds.), Stability Problems for Stochastic Models. Proceedings, 1991. VIII, 229 pages. 1993.

Vol. 1547: P. Harmand, D. Werner, W. Werner, M-ideals in Banach Spaces and Banach Algebras. VIII, 387 pages. 1993.

Vol. 1548: T. Urabe, Dynkin Graphs and Quadrilateral Singularities. VI, 233 pages. 1993.

Vol. 1549: G. Vainikko, Multidimensional Weakly Singular Integral Equations. XI, 159 pages. 1993.

Vol. 1550: A. A. Gonchar, E. B. Saff (Eds.), Methods of Approximation Theory in Complex Analysis and Mathematical Physics IV, 222 pages, 1993.

Vol. 1551: L. Arkeryd, P. L. Lions, P.A. Markowich, S.R. S. Varadhan. Nonequilibrium Problems in Many-Particle Systems. Montecatini, 1992. Editors: C. Cercignani, M. Pulvirenti. VII, 158 pages 1993.

Vol. 1552: J. Hilgert, K.-H. Neeb, Lie Semigroups and their Applications. XII, 315 pages. 1993.

Vol. 1553: J.-L- Colliot-Thélène, J. Kato, P. Vojta. Arithmetic Algebraic Geometry. Editor: E. Ballico. VII, 223 pages. 1993.

Vol. 1554: A. K. Lenstra, H. W. Lenstra, Jr. (Eds.), The Development of the Number Field Sieve. VIII, 131 pages. 1993.

Vol. 1555: O. Liess, Conical Refraction and Higher Microlocalization. X, 389 pages. 1993.

Vol. 1556: S. B. Kuksin, Nearly Integrable Infinite-Dimensional Hamiltonian Systems. XXVII, 101 pages. 1993.

Vol. 1557: J. Azéma, P. A. Meyer, M. Yor (Eds.), Séminaire de Probabilités XXVII. VI, 327 pages. 1993.

Vol. 1558: T. J. Bridges, J. E. Furter, Singularity Theory and Equivariant Symplectic Maps. VI, 226 pages. 1993.

Vol. 1559: V. G. Sprindžuk, Classical Diophantine Equations. XII, 228 pages. 1993.

Vol. 1560: T. Bartsch, Topological Methods for Variational Problems with Symmetries. X, 152 pages. 1993.

Vol. 1561: I. S. Molchanov, Limit Theorems for Unions of Random Closed Sets. X, 157 pages. 1993.

Vol. 1562: G. Harder, Eisensteinkohomologie und die Konstruktion gemischter Motive. XX, 184 pages. 1993.

Vol. 1563: E. Fabes, M. Fukushima, L. Gross, C. Kenig, M. Röckner, D. W. Stroock, Dirichlet Forms. Varenna, 1992. Editors: G. Dell'Antonio, U. Mosco. VII, 245 pages. 1993.

Vol. 1564: J. Jorgenson, S. Lang, Basic Analysis of Regularized Series and Products. IX, 122 pages. 1993.

Vol. 1565: L. Boutet de Monvel, C. De Concini, C. Procesi, P. Schapira, M. Vergne. D-modules, Representation Theory, and Quantum Groups. Venezia, 1992. Editors: G. Zampieri, A. D'Angelo. VII, 217 pages. 1993.